CAMBRIDGE LIBRARY

Books of enduring scholarly value

Life Sciences

Until the nineteenth century, the various subjects now known as the life sciences were regarded either as arcane studies which had little impact on ordinary daily life, or as a genteel hobby for the leisured classes. The increasing academic rigour and systematisation brought to the study of botany, zoology and other disciplines, and their adoption in university curricula, are reflected in the books reissued in this series.

Life and Letters of Thomas Henry Huxley

Thomas Henry Huxley (1825–95), the English biologist and naturalist, was known as 'Darwin's Bulldog', and is best remembered today for his vociferous support for Darwin's theory of evolution. He was, however, an influential naturalist, anatomist and religious thinker, who coined the term 'agnostic' to describe his own beliefs. Almost entirely self-educated, he became an authority in anatomy and palaeontology, and after the discovery of the archaeopteryx, he was the first to suggest that birds had evolved from dinosaurs. He was also a keen promoter of scientific education who strove to make science a paid profession, not dependent on patronage or wealth. Published in 1903, this three-volume work, compiled by his son Leonard Huxley, is the second and most complete edition of Huxley's biography and selected letters. Volume 1 covers the period 1825–69, including his expedition to Australasia and the publication of the *On the Origin of Species* (1859).

Cambridge University Press has long been a pioneer in the reissuing of out-of-print titles from its own backlist, producing digital reprints of books that are still sought after by scholars and students but could not be reprinted economically using traditional technology. The Cambridge Library Collection extends this activity to a wider range of books which are still of importance to researchers and professionals, either for the source material they contain, or as landmarks in the history of their academic discipline.

Drawing from the world-renowned collections in the Cambridge University Library, and guided by the advice of experts in each subject area, Cambridge University Press is using state-of-the-art scanning machines in its own Printing House to capture the content of each book selected for inclusion. The files are processed to give a consistently clear, crisp image, and the books finished to the high quality standard for which the Press is recognised around the world. The latest print-on-demand technology ensures that the books will remain available indefinitely, and that orders for single or multiple copies can quickly be supplied.

The Cambridge Library Collection brings back to life books of enduring scholarly value (including out-of-copyright works originally issued by other publishers) across a wide range of disciplines in the humanities and social sciences and in science and technology.

Life and Letters of
Thomas Henry Huxley

VOLUME 1

LEONARD HUXLEY

CAMBRIDGE UNIVERSITY PRESS

Cambridge, New York, Melbourne, Madrid, Cape Town,
Singapore, São Paolo, Delhi, Tokyo, Mexico City

Published in the United States of America by Cambridge University Press, New York

www.cambridge.org
Information on this title: www.cambridge.org/9781108040457

© in this compilation Cambridge University Press 2012

This edition first published 1903
This digitally printed version 2012

ISBN 978-1-108-04045-7 Paperback

LIFE AND LETTERS

OF

THOMAS HENRY HUXLEY

Walker & Boutall ph. sc.

T. H. Huxley
From a Daguerreotype made in 1846.

Life and Letters

OF

Thomas Henry Huxley

BY HIS SON

LEONARD HUXLEY

IN THREE VOLUMES

VOL. I

London

MACMILLAN AND CO., Limited

NEW YORK: THE MACMILLAN COMPANY

1903

First Edition, 2 vols. 8vo, October 1900.
Reprinted November and December 1900.
Second Edition, Globe 8vo (*Eversley Series,* 3 *vols.*), 1903.

PREFACE TO THE SECOND EDITION

My thanks are due to a number of friends for pointing out to me various misprints and other errors or omissions which had passed unnoticed in the first edition. Professor Howes in particular has furnished the titles of several scientific memoirs, the identification of which is due to his careful research in the journals of the learned societies.

Of the fresh material which has come into my hands recently, I have printed two or three letters. In one or two passages, also, I have altered the wording slightly in deference to Mr. Herbert Spencer, who thought that despite the definite statement quoted from a letter of his on ii. 442, the public would receive the impression that my father's reading of his proofs had extended to all his works. As a matter of fact, to quote the substance of my subsequent letter to the *Athenæum*, that reading was restricted to Mr. Spencer's biological writings, the

First Principles, and two other fragments amounting to thirty-two pages. With characteristic accuracy, moreover, Mr. Spencer tells me that comparison of the proofs of the three latter with the published works shows the precise number of the "brilliant speculations choked in an embryonic state" therein by his "devil's advocacy" (ii. 18) to be four. But the period assigned for this "devil's advocacy," going back "thirty odd years," from 1884 to the beginning of my father's acquaintance with Mr. Spencer, indicates that the playful allusion must be as much to the informal dialectics of conversation as to serious written work, for the reading of proofs referred to above, only began with the *Synthetic Philosophy* in 1860.

L. H.

November 1902.

PREFACE TO THE FIRST EDITION

My father's life was one of so many interests, and his work was at all times so diversified, that to follow each thread separately, as if he had been engaged on that alone for a time, would be to give a false impression of his activity and the peculiar character of his labours. All through his active career he was equally busy with research into nature, with studies in philosophy, with teaching and administrative work. The real measure of his energy can only be found when all these are considered together. Without this there can be no conception of the limitations imposed upon him in his chosen life's work. The mere amount of his research is greatly magnified by the smallness of the time allowed for it.

But great as was the impression left by these researches in purely scientific circles, it is not by them alone that he made his impression upon the mass of his contemporaries. They were chiefly moved by something over and above his wide knowledge in so many fields,—by his passionate sincerity, his interest not only in pure knowledge, but in human life, by his belief that the interpretation of the book of nature was not to be kept apart from the ultimate problems of existence; by the love of truth, in short, both theoretical and practical, which gave the key to the character of the man himself.

Accordingly, I have not discussed with any fulness the value of his technical contributions to natural science; I have not drawn up a compendium of his philosophical views. One is a work for specialists; the other can be gathered from his published works. I have endeavoured rather to give the public a picture, so far as I can, of the man himself, of his aims in the many struggles in which he was engaged, of his character and temperament, and the circumstances under which his various works were begun and completed.

So far as possible, I have made his letters, or extracts from them, tell the story of his life. If those of any given period are diverse in tone and character, it is simply because they reflect an equal diversity of occupations and interests. Few of the letters, however, are of any great length; many are little more than hurried notes; others, mainly of private interest, supply a sentence here and there to fill in the general outline.

Moreover, whenever circumstances permit, I have endeavoured to make my own part in the book entirely impersonal. My experience is that the constant iteration by the biographer of his relationship to the subject of his memoir can become exasperating to the reader; so that, at the risk of offending in the opposite direction, I have chosen the other course.

Lastly, I have to express my grateful thanks to all who have sent me letters or supplied information, and especially to Dr. J. H. Gladstone, Sir Mountstuart Grant Duff, Professor Howes, Professor Henry Sidgwick, and Sir Spencer Walpole, for their contributions to the book; but above all to Sir Joseph Hooker and Sir Michael Foster, whose invaluable help in reading proofs and making suggestions has been, as it were, a final labour of love for the memory of their old friend.

L. H.

CONTENTS

PORTRAIT

T. H. HUXLEY

From a Daguerreotype made in 1846.

Frontispiece.

CHAPTER I

1825–1842

IN the year 1825 Ealing was as quiet a country village as could be found within a dozen miles of Hyde Park Corner. Here stood a large semi-public school, which had risen to the front rank in numbers and reputation under Dr. Nicholas, of Wadham College, Oxford, who in 1791 became the son-in-law and successor of the previous master.

The senior assistant-master in this school was George Huxley, a tall, dark, rather full-faced man, quick tempered, and distinguished, in his son's words, by "that glorious firmness which one's enemies called obstinacy." In the year 1810 he had married Rachel Withers; she bore five sons and three daughters, of whom one son and one daughter died in infancy; the seventh and youngest surviving child was Thomas Henry.

George Huxley, the master at Ealing, was the second son of Thomas Huxley and Margaret James, who were married at St. Michael's, Coventry, on September 8, 1773. This Thomas Huxley continued to live at Coventry until his death in January 1796, when he left behind him a large family and no very great wealth. The most notable item in the latter is the "capital Messuage, by me lately purchased of Mrs. Ann Thomas," which he directs to be sold to pay his debts—an inn, apparently, for the testator is described as a victualler. Family tradition tells that he came to Coventry from Lichfield, and if so, he and his sons after him exemplify the tendency to move south, which is to be observed in those of the same name who migrated from their original home in Cheshire. This home is represented to-day by a farm in Broxton Hundred, about eight miles from Chester, called Huxley Hall. From this centre Huxleys spread to the neighbouring villages, such as Overton and Eccleston, Clotton and Duddon, Tattenhall and Wettenhall; others to Chester and Brindley near Nantwich. The southward movement carries some to the Welsh border, others into Shropshire. The Wettenhall family established themselves in the fourth generation at Rushall, and held property in Handsworth and Walsall; the Brindley family sent a branch to Macclesfield, whose representative, Samuel, must have been on the town council when the Young Pretender rode through on his way to Derby, for he was mayor in 1746; while at the end of the sixteenth century, George, the disinherited heir of Brindley, became a merchant in London, and purchased Wyre Hall at Edmonton, where his descendants lived for four generations, his grandson being knighted by Charles II. in 1663.

But my father had no particular interest in tracing his early ancestry. "My own genealogical inquiries," he said, "have taken me so far back that I confess the later stages do not interest me." Towards the end of his life, however, my mother persuaded him to see what could be found out about Huxley Hall and the origin of the name. This proved to be from the manor of Huxley or Hodesleia, whereof one Swanus de Hockenhull was enfeoffed by the abbot and convent of St. Werburgh in the time of Richard I. Of the grandsons of this Swanus, the eldest kept the manor and name of Hockenhull (which is still extant in the Midlands); the younger ones took their name from the other fief.

But the historian of Cheshire records the fact that owing to the respectability of the name, it was unlawfully assumed by divers "losels and lewd fellows of the baser sort," and my father, with a fine show of earnestness, used to declare that he was certain the legitimate owners of the name were far too sober and respectable to have produced such a reprobate as himself, and one of these "losels" must be his progenitor.

Thomas Henry Huxley was born at Ealing on May 4, 1825, "about eight o'clock in the morning."[1] "I am not aware," he tells us playfully in his Auto-

[1] So in the Autobiography, but 9.30 according to the family Bible.

biography, "that any portents preceded my arrival in this world, but, in my childhood, I remember hearing a traditional account of the manner in which I lost the chance of an endowment of great practical value. The windows of my mother's room were open, in consequence of the unusual warmth of the weather. For the same reason, probably, a neighbouring bee-hive had swarmed, and the new colony, pitching on the window-sill, was making its way into the room when the horrified nurse shut down the sash. If that well-meaning woman had only abstained from her ill-timed interference, the swarm might have settled on my lips, and I should have been endowed with that mellifluous eloquence which, in this country, leads far more surely than worth, capacity, or honest work, to the highest places in Church and State. But the opportunity was lost, and I have been obliged to content myself through life with saying what I mean in the plainest of plain language, than which, I suppose, there is no habit more ruinous to a man's prospects of advancement."

As to his debt, physical and mental, to either parent, he writes as follows :—

Physically I am the son of my mother so completely— even down to peculiar movements of the hands, which made their appearance in me as I reached the age she had when I noticed them—that I can hardly find any trace of my father in myself, except an inborn faculty for drawing, which, unfortunately, in my case, has never been cultivated, a hot temper, and that amount of tenacity of purpose which unfriendly observers sometimes call obstinacy.

My mother was a slender brunette, of an emotional and energetic temperament, and possessed of the most piercing black eyes I ever saw in a woman's head. With no more education than other women of the middle classes in her day, she had an excellent mental capacity. Her most distinguishing characteristic, however, was rapidity of thought. If one ventured to suggest that she had not taken much time to arrive at any conclusion, she would say, "I cannot help it; things flash across me." That peculiarity has been passed on to me in full strength; it has often stood me in good stead; it has sometimes played me sad tricks, and it has always been a danger. But, after all, if my time were to come over again, there is nothing I would less willingly part with than my inheritance of mother-wit.

Restless, talkative, untiring to the day of her death, she was at sixty-six "as active and energetic as a young woman." His early devotion to her was remarkable. Describing her to his future wife, he writes :—

As a child my love for her was a passion. I have lain awake for hours crying because I had a morbid fear of her death ; her approbation was my greatest reward, her displeasure my greatest punishment.

I have next to nothing to say about my childhood (he continues in the Autobiography). In later years my mother, looking at me almost reproachfully, would sometimes say, "Ah . you were such a pretty boy !" whence I had no difficulty in concluding that I had not fulfilled my early promise in the matter of looks. In fact, I have a distinct recollection of certain curls of which I was vain, and of a conviction that I closely resembled that handsome, courtly gentleman, Sir Herbert Oakley, who was vicar of our parish, and who was as a god to us country folk, because he was occasionally visited by the then

Prince George of Cambridge. I remember turning my pinafore wrong side forwards in order to represent a surplice, and preaching to my mother's maids in the kitchen as nearly as possible in Sir Herbert's manner one Sunday morning when the rest of the family were at church. That is the earliest indication of the strong clerical affinities which my friend Mr. Herbert Spencer has always ascribed to me, though I fancy they have for the most part remained in a latent state.

There remains no record of his having been a very precocious child. Indeed, it is usually the eldest child whose necessary companionship with his elders wins him this reputation. The youngest remains a child among children longer than any other of his brothers and sisters.

One talent, however, displayed itself early. The faculty of drawing he inherited from his father. But on the queer principle that training is either unnecessary to natural capacity or even ruins it, he never received regular instruction in drawing; and his draughtsmanship, vigorous as it was, and a genuine medium of artistic expression as well as an admirable instrument in his own especial work, never reached the technical perfection of which it was naturally capable.

The amount of instruction, indeed, of any kind which he received was scanty in the extreme. For a couple of years, from the age of eight to ten, he was given a taste of the unreformed public school life, where, apart from the rough and ready mode of instruction in vogue and the necessary obedience enforced to certain rules, no means were taken to reach the boys themselves, to guide them and help them in their school life. The new-comer was left to struggle for himself in a community composed of human beings at their most heartlessly cruel age, untempered by any external influence.

Here he had little enough of mental discipline, or that deliberate training of character which is a leading object of modern education. On the contrary, what he learnt was a knowledge of undisciplined human nature.

My regular school training (he tells us) was of the briefest, perhaps fortunately ; for though my way of life has made me acquainted with all sorts and conditions of men, from the highest to the lowest, I deliberately affirm that the society I fell into at school was the worst I have ever known. We boys were average lads, with much the same inherent capacity for good and evil as any others ; but the people who were set over us cared about as much for our intellectual and moral welfare as if they were baby-farmers. We were left to the operation of the struggle for existence among ourselves ; bullying was the least of the ill practices current among us. Almost the only cheerful reminiscence in connection with the place which arises in my mind is that of a battle I had with one of my classmates, who had bullied me until I could stand it no longer. I was a very slight lad, but there was a wild-cat element in me which, when roused, made up for lack of weight, and I licked my adversary effectually. However, one of my first experiences of the extremely rough-and-ready nature of justice, as exhibited by the course of things in general, arose out of the fact that I—the victor—had a black eye, while he—the vanquished—had none, so that I got into disgrace and he did not. We made it up, and thereafter I was un-molested. One of the greatest shocks I ever received in my life was to be told a dozen years afterwards by the groom who brought me my horse in a stable-yard in Sydney that he was my quondam antagonist. He had a long story of family misfortune to account for his position ; but at that time it was necessary to deal very cautiously with mysterious strangers in New South Wales, and on inquiry I found that the unfortunate young man had not only been "sent out," but had undergone more than one colonial conviction.

His brief school career was happily cut short by the break-up of the Ealing establishment. On the death of Dr. Nicholas, his sons attempted to carry on the school; but the numbers declined rapidly, and George Huxley, about 1835, returned to his native town of Coventry, where he obtained the modest post of manager of the Coventry savings bank, while his daughters eked out the slender family resources by keeping school.

In the meantime the boy Tom, as he was usually called, got little or no regular instruction. But he had an inquiring mind, and a singularly early turn for metaphysical speculation. He read everything he could lay hands on in his father's library. Not satisfied with the ordinary length of the day, he used, when a boy of twelve, to light his candle before dawn, pin a blanket round his shoulders, and sit up in bed to read Hutton's *Geology*. He discussed all manner of questions with his parents and friends, for his quick and eager mind made it possible for him to have friendships with people considerably older than

himself. Among these may especially be noted his
medical brother-in-law, Dr. Cooke of Coventry, who
had married his sister Ellen in 1839, and through
whom he early became interested in human anatomy;
and George Anderson May, at that time in business
at Hinckley (a small weaving centre some dozen
miles distant from Coventry), whom his friends who
knew him afterwards in the home which he made
for himself on the farm at Elford, near Tamworth,
will remember for his genial spirit and native love of
letters. There was a real friendship between the
two. The boy of fifteen notes down with pleasure
his visits to the man of six-and-twenty, with whom
he could talk freely of the books he read, and the
ideas he gathered about philosophy.

Afterwards, however, their ways lay far apart,
and I believe they did not meet again until the
seventies, when Mr. May sent his children to be educated in London, and his youngest son was at school
with me; his younger daughter studied art at the
Slade School with my sisters, and both found a warm
welcome in the home circle at Marlborough Place.

One of his boyish speculations was as to what
would become of things if their qualities were taken
away; and lighting upon Sir William Hamilton's
Logic, he devoured it to such good effect that when,
years afterwards, he came to tackle the greater
philosophers, especially the English and the German,
he found he had already a clear notion of where the
key of metaphysic lay.

This early interest in metaphysics was another form of the intense curiosity to discover the motive principle of things, the why and how they act, that appeared in the boy's love of engineering and of anatomy. The unity of this motive and the accident which bade fair to ruin his life at the outset, and actually levied a lifelong tax upon his bodily vigour, are best told in his own words :—

As I grew older, my great desire was to be a mechanical engineer, but the fates were against this, and while very young I commenced the study of medicine under a medical brother-in-law. But, though the Institute of Mechanical Engineers would certainly not own me, I am not sure that I have not all along been a sort of mechanical engineer *in partibus infidelium.* I am now occasionally horrified to think how little I ever knew or cared about medicine as the art of healing. The only part of my professional course which really and deeply interested me was physiology, which is the mechanical engineering of living machines ; and, notwithstanding that natural science has been my proper business, I am afraid there is very little of the genuine naturalist in me. I never collected anything, and species work was always a burden to me ; what I cared for was the architectural and engineering part of the business, the working out the wonderful unity of plan in the thousands and thousands of diverse living constructions, and the modifications of similar apparatuses to serve diverse ends. The extraordinary attraction I felt towards the study of the intricacies of living structure nearly proved fatal to me at the outset. I was a mere boy—I think between thirteen and fourteen years of age—when I was taken by some older student friends of mine to the first *post-mortem* examination I ever attended. All my life I have been most unfortunately sensitive to the disagreeables which

attend anatomical pursuits, but on this occasion my
curiosity overpowered all other feelings, and I spent two
or three hours in gratifying it. I did not cut myself,
and none of the ordinary symptoms of dissection-poison
supervened, but poisoned I was somehow, and I remember
sinking into a strange state of apathy. By way of a last
chance, I was sent to the care of some good, kind people,
friends of my father's, who lived in a farmhouse in the
heart of Warwickshire. I remember staggering from my
bed to the window on the bright spring morning after
my arrival, and throwing open the casement. Life seemed
to come back on the wings of the breeze, and to this day
the faint odour of wood-smoke, like that which floated
across the farmyard in the early morning, is as good
to me as the "sweet south upon a bed of violets." I
soon recovered, but for years I suffered from occasional
paroxysms of internal pain, and from that time my
constant friend, hypochondriacal dyspepsia, commenced
his half-century of co-tenancy of my fleshly tabernacle.

Some little time after his return from the voyage
of the *Rattlesnake*, Huxley succeeded in tracing his
good Warwickshire friends again. A letter of May
11, 1852, from one of them, Miss K. Jaggard, tells
how they had lost sight of the Huxleys after their
departure from Coventry ; how they were themselves
dispersed by death, marriage, or retirement; and
then proceeds to draw a lively sketch of the long
delicate-looking lad, which clearly refers to this
period or a little later.

My brother and sister who were living at Grove Fields
when you visited there, have now retired from the cares
of business, and are living very comfortably at Leam-

ington. . . . I suppose you remember Mr. Joseph Russell, who used to live at Avon Dassett. He is now married and gone to live at Grove Fields, so that it is still occupied by a person of the same name as when you knew it. But it is very much altered in appearance since the time when such merry and joyous parties of aunts and cousins used to assemble there. I assure you we have often talked of "Tom Huxley" (who was sometimes one of the party), looking so thin and ill, and pretending to make hay with one hand, while in the other he held a German book! Do you remember it? And the picnic at Scar Bank? And how often too your patience was put to the test in looking for your German books which had been hidden by some of those playful companions who were rather less inclined for learning than yourself?

It is interesting to see from this letter and from a journal, to be quoted hereafter, that he had thus early begun to teach himself German, an under-taking more momentous in its consequences than the boy dreamed of. The knowledge of German thus early acquired was soon of the utmost service in making him acquainted with the advance of biological investigation on the Continent at a time when few indeed among English men of science were able to follow it at first hand, and turn the light of the newest theories upon their own researches.

It is therefore peculiarly interesting to note the cause which determined the young Huxley to take up the study of so little read a language. I have more than once heard him say that this was one half of the debt he owed to Carlyle, the other half being an intense hatred of shams of every sort and kind.

The translations from the German, the constant references to German literature and philosophy, fired him
to try the vast original from which these specimens
were quarried, for the sake partly of the literature,
but still more of the philosophy. The translation
of *Wilhelm Meister*, and some of the *Miscellaneous
Essays*, together with *The French Revolution*, were
certainly among works of Carlyle with which he first
made acquaintance, to be followed later by *Sartor
Resartus*, which for many years afterwards was his
Enchiridion, as he puts it in an unpublished auto-
biographical fragment.

By great good fortune, a singularly interesting
glimpse of my father's life from the age of fifteen onwards has been preserved in the shape of a fragmentary journal which he entitled, German fashion,
Thoughts and Doings. Begun on September 29, 1840,
it is continued for a couple of years, and concludes
with some vigorous annotations in 1845, when the
little booklet emerged from a three years' oblivion at
the bottom of an old desk. Early as this journal is, in
it the boy displays three habits afterwards characteristic of the man : the habit of noting down any striking thought or saying he came across in the course of
his reading ; of speculating on the causes of things
and discussing the right and wrong of existing
institutions ; and of making scientific experiments,
using them to correct his theories.

The first entry, the heading, as it were, and keynote of all the rest, is a quotation from Novalis :—

"Philosophy can bake no bread ; but it can prove for us God, freedom, and immortality. Which, now, is more practical, Philosophy or Economy?" The reference here given is to a German edition of Novalis, so that it seems highly probable that the boy had learnt enough of the language to translate a bit for himself, though, as appears from entries in 1841, he had still to master the grammar completely.

In science, he was much interested in electricity ; he makes a galvanic battery "in view of experiment to get crystallised carbon. Got it deposited, but not crystallised." Other experiments and theorising upon them are recorded in the following year. Another entry showing the courage of youth, deserves mention :—

"*Oct.* 5 (1840).—Began speculating on the cause of colours at sunset. Has any explanation of them ever been attempted?" which is supplemented by an extract "from old book."

We may also remark the early note of Radicalism and resistance to anything savouring of injustice or oppression, together with the naïve honesty of the admission that his opinions may change with years.

Oct. 25 (at Hinckley).—Read Dr. S. Smith on the Divine Government.—Agree with him partly.—I should say that a general belief in his doctrines would have a very injurious effect on morals.

Nov. 22.—. . . Had a long talk with my mother and father about the right to make Dissenters pay church rates —and whether there ought to be any Establishment. I maintain that there ought not in both cases—I wonder what will be my opinion ten years hence? I think now that it is against all laws of justice to force men to support a church with whose opinions they cannot conscientiously agree. The argument that the rate is so small is very fallacious. It is as much a sacrifice of principle to do a little wrong as to do a great one.

Nov. 22 (Hinckley).—Had a long argument with Mr. May on the nature of the soul and the difference between it and matter. I maintained that it could not be proved that matter is *essentially*—as to its base—different from soul. Mr. M. wittily said, soul was the perspiration of matter.

We cannot find the absolute basis of matter : we only know it by its properties ; neither know we the soul in any other way. *Cogito ergo sum* is the only thing that we *certainly* know.

Why may not soul and matter be of the same substance (*i.e.* basis whereon to fix qualities, for we cannot suppose a quality to exist *per se*—it must have a something to qualify), but with different qualities.

Let us suppose then an Eon—a something with no quality but that of existence—this Eon endued with all the intelligence, mental qualities, and that in the highest degree—is God. This combination of intelligence with existence we may suppose to have existed from eternity. At the creation we may suppose that a portion of the Eon was separated from the intelligence, and it was ordained —it became a natural law—that it should have the properties of gravitation, etc.—that is, that it should give to man the ideas of those properties. The Eon in this state is matter in the abstract. Matter, then, is Eon in the simplest form in which it possesses qualities appreciable by the senses. Out of this matter, by the superimposition of fresh qualities, was made all things that are.

1841

Jan. 7.—Came to Rotherhithe.[1]

June 20.—What have I done in the way of acquiring knowledge since January?

Projects begun—
1. German ⎱ to be learnt.
2. Italian ⎰
3. To read Müller's *Physiology*,
4. To prepare for the Matriculation Examination at London University which requires knowledge of :—

    (*a*) Algebra—Geometry ⎱ did not begin to read
    (*b*) Natural Philosophy ⎰   for this till April.
    (*c*) Chemistry.
    (*d*) Greek—Latin.
    (*e*) English History down to end of seventeenth century.
    (*f*) Ancient History.
        English Grammar.

5. To make copious notes of all things I read.

Projects completed—
1. Partly. 2. Not at all. 3 and 5, stuck to these pretty closely.
4. (*e*) Read as far as Henry III. in Hume.
    (*a*) Evolution and involution.
    (*b*) Refraction of light—Polarisation partly.
    (*c*) Laws of combination—must read them over again.
    (*d*) Nothing.
    (*f*) Nothing.

I must get on faster than this. I *must* adopt a fixed plan of studies, for unless this is done I find time slips

---

[1] See Chap. II.

away without knowing it—and let me remember this—
that it is better to read a little and thoroughly, than
cram a crude undigested mass into my head, though it be
great in quantity.

(This is about the only resolution I have ever stuck
to—1845.)

[Well do I remember how in that little narrow surgery
I used to work morning after morning and evening after
evening at that insufferably dry and profitless book,
Hume's *History*, how I worked against hope through the
series of thefts, robberies, and throat-cutting in those
three first volumes, and how at length I gave up the
task in utter disgust and despair.

Mackintosh's *History*, on the other hand, I remember
reading with great pleasure, and also Guizot's *Civilisation
in Europe;* the scientific theoretical form of the latter
especially pleased me, but the want of sufficient knowledge to test his conclusions was a great drawback. 1845]

There follow notes of work done in successive
weeks—June 20 to August 9, and September 27 to
October 4. History, German, Mathematics, Physics,
Physiology ; makes an electro-magnet ; reads Guizot's
History of Civilisation in Europe, on which he remarks
" an excellent work—very tough reading, though."

At the beginning of October, under " Miscellaneous," "Became acquainted with constitution of
French Chambre des députés and their parties."

It was his practice to note any sayings that struck
him :—

Truths : " I hate all people who want to found sects.
It is not error but sects—it is not error but sectarian
error, nay, and even sectarian truth, which causes the
unhappiness of mankind."—Lessing.

"It is only necessary to grow old to become more indulgent. I see no fault committed that I have not committed myself. . . ."—Goethe.

" One solitary philosopher may be great, virtuous, and happy in the midst of poverty, but not a whole nation. . . ."—Isaac Iselin.

1842

Jan. 30, *Sunday evening.*—I have for some time been pondering over a classification of knowledge. My scheme is to divide all knowledge in the first place into two grand divisions. 1. Objective—that for which a man is indebted to the external world ; and 2. Subjective— that which he has acquired or may acquire by inward contemplation.

Metaphysics comes immediately, of course, under the first (2) head—that is to say, the relations of the mind to itself ; of this Mathematics and Logic, together with Theology, are branches.

I am in doubt under which head to put morality, for I cannot determine exactly in my own mind whether morality can exist independent of others, whether the idea of morality could ever have arisen in the mind of an isolated being or not. I am rather inclined to the opinion that it is objective.

Under the head of objective knowledge comes first Physics, including the whole body of the relations of inanimate unorganised bodies ; secondly, Physiology. Including the structure and functions of animal bodies, including language and Psychology ; thirdly comes History.

One object for which I have attempted to form an arrangement of knowledge is that I may test the amount of my own acquirements. I shall form an extensive list of subjects on this plan, and as I acquire any one of them I shall strike it out of the list. May the list soon get black ! though at present I shall hardly be able, I am afraid, to spot the paper.

(A prophecy ! a prophecy, 1845 !)

April 1842 introduces a number of quotations from Carlyle's Miscellaneous Writings, "Characteristics," some clear and crisp, others sinking into Carlyle's own vein of speculative mysticism, *e.g.*

"In the mind as in the body the sign of health is unconsciousness."

"Of our thinking it is but the upper surface that we shape into articulate thought ; underneath the region of argument and conscious discourse lies the region of meditation."

"Genius is ever a secret to itself."

"The healthy understanding, we should say, is neither the argumentative nor the Logical, but the Intuitive, for the end of understanding is not to prove and find reasons but to know and believe " (!)

"The ages of heroism are not ages of Moral Philosophy Virtue, when it is philosophised of, has become aware of itself, is sickly and beginning to decline."

At the same time more electrical experiments are recorded ; and theories are advanced with pros and cons to account for the facts observed.

The last entry was made three years later—

Oct. 1845.—I have found singular pleasure—having accidentally raked this Büchlein from a corner of my desk—in looking over these scraps of notices of my past existence ; an illustration of J. Paul's saying that a man has but to write down his yesterday's doings, and forthwith they appear surrounded with a poetic halo.

But after all, these are but the top skimmings of these five years' living. I hardly care to look back into the seething depths of the working and boiling mass that lay beneath all this froth, and indeed I hardly know whether I could give myself any clear account of it. Remembrances of physical and mental pain . . . absence of sympathy, and thence a choking up of such few ideas as I did form clearly within my own mind.

Grief too, yet at the misfortune of others, for I have had few properly my own ; so much the worse, for in that case I might have said or done somewhat, but here was powerless.

Oh, Tom, trouble not thyself about sympathy ; thou hast two stout legs and young, wherefore need a staff ?

Furthermore, it is twenty minutes past two, and time to go to bed.

Büchlein, it will be long before my secretiveness remains so quiet again ; make the most of what thou hast got.

CHAPTER II

1841–1846

THE migration to Rotherhithe, noted under date of January 9, 1841, was a fresh step in his career. In 1839 both his sisters married, and both married doctors. Dr. Cooke, the husband of the elder sister, who was settled in Coventry, had begun to give him some instruction in the principles of medicine as early as the preceding June. It was now arranged that he should go as assistant to Mr. Chandler, of Rotherhithe, a practical preliminary to walking the hospitals and obtaining a medical degree in London. His experiences among the poor in the dock region of the East of London—for Dr. Chandler had charge of the parish—supplied him with a grim commentary on his diligent reading in Carlyle. Looking back on this period, he writes :—

The last recorded speech of Professor Teufelsdröckh proposes the toast 'Die Sache der Armen in Gottes und Teufels-namen' (The cause of the Poor in Heaven's name and ——'s). The cause of the Poor is the burden of *Past and Present, Chartism,* and *Latter-Day Pamphlets.*

To me . . . this advocacy of the cause of the poor
appealed very strongly . . . because . . . I had had the
opportunity of seeing for myself something of the way
the poor live. Not much, indeed, but still enough to
give a terrible foundation of real knowledge to my
speculations.

After telling how he came to know something of
the East End, he proceeds :—

I saw strange things there—among the rest, people
who came to me for medical aid, and who were really
suffering from nothing but slow starvation. I have not
forgotten—am not likely to forget so long as memory
holds—a visit to a sick girl in a wretched garret where
two or three other women, one a deformed woman, sister
of my patient, were busy shirt-making. After due
examination, even my small medical knowledge sufficed
to show that my patient was merely in want of some
better food than the bread and bad tea on which these
people were living. I said so as gently as I could, and
the sister turned upon me with a kind of choking passion.
Pulling out of her pocket a few pence and halfpence, and
holding them out, "That is all I get for six-and-thirty
hours' work, and you talk about giving her proper food."

Well, I left that to pursue my medical studies, and it
so happened the shortest way between the school which I
attended and the library of the College of Surgeons, where
my spare hours were largely spent, lay through certain
courts and alleys, Vinegar Yard and others, which are
now nothing like what they were then. Nobody would
have found robbing me a profitable employment in those
days, and I used to walk through these wretched dens
without let or hindrance. Alleys nine or ten feet wide,
I suppose, with tall houses full of squalid drunken men
and women, and the pavement strewed with still more
squalid children. The place of air was taken by a steam

of filthy exhalations; and the only relief to the general
dull apathy was a roar of words—filthy and brutal beyond
imagination — between the close-packed neighbours,
occasionally ending in a general row. All this almost
within hearing of the traffic of the Strand, within easy
reach of the wealth and plenty of the city.

I used to wonder sometimes why these people did not
sally forth in mass and get a few hours' eating and
drinking and plunder to their hearts' content, before the
police could stop and hang a few of them. But the poor
wretches had not the heart even for that. As a slight,
wiry Liverpool detective once said to me when I asked
him how it was he managed to deal with such hulking
ruffians as we were among, "Lord bless you, sir, drink
and disease leave nothing in them."

This early contact with the sternest facts of the
social problem impressed him profoundly. And
though not actively employed in what is generally
called "philanthropy," still he did his part, hopefully
but soberly, not only to throw light on the true
issues and to strip away make-believe from them, but
also to bring knowledge to the working classes, and
to institute machinery by which capacity should be
caught and led to a position where it might be useful
instead of dangerous to social order.

After some time, however, he left Mr. Chandler
to join his second brother-in-law,[1] who had set up in
the north of London, and to whom he was duly
apprenticed, as his brother James had been before
him. This change gave him more time and opportunity to pursue his medical education. He attended

---

[1] John Godwin Scott.

lectures at the Sydenham College, and, as has been seen, began to prepare for the matriculation examination of the University of London. At the Sydenham College he met with no little success, winning, besides certificates of merit in other departments, a prize—his first prize—for botany. His vivid recollections, given below, of this entry into the scientific arena are taken from a journal he kept for his fiancée during his absence from Sydney on the cruises of the *Rattlesnake*.

ON BOARD H.M.S. *RATTLESNAKE*, CHRISTMAS 1847.

Next summer it will be six years since I made my first trial in the world. My first public competition, small as it was, was an epoch in my life. I had been attending (it was my first summer session) the botanical lectures at Chelsea. One morning I observed a notice stuck up—a notice of a public competition for medals, etc., to take place on the 1st August (if I recollect right). It was then the end of May or thereabouts. I remember looking longingly at the notice, and some one said to me, "Why don't you go in and try for it?" I laughed at the idea, for I was very young, and my knowledge somewhat of the vaguest. Nevertheless I mentioned the matter to S.[1] when I returned home. He likewise advised me to try, and so I determined I would. I set to work in earnest, and perseveringly applied myself to such works as I could lay my hands on, Lindley's and Decandolle's *Systems* and the *Annales des Sciences Naturelles* in the British Museum. I tried to read Schleiden, but my German was insufficient.

For a young hand I worked really hard from eight

---

[1] His brother-in-law.

or nine in the morning until twelve at night, besides a
long hot summer's walk over to Chelsea two or three
times a week to hear Lindley.   A great part of the time
I worked till sunrise. The result was a sort of
ophthalmia which kept me from reading at night for
months afterwards.

The day of examination came, and as I went along
the passage to go out I well remember dear Lizzie,[1] half
in jest, half in earnest, throwing her shoe after me, as
she said, for luck. She was alone, beside S., in the
secret, and almost as anxious as I was. How I reached
the examination room I hardly know, but I recollect
finding myself at last with pen and ink and paper before
me and five other beings, all older than myself, at a
long table. We stared at one another like strange cats
in a garret, but at length the examiner (Ward) entered,
and before each was placed the paper of questions and
sundry plants. I looked at my questions, but for some
moments could hardly hold my pen, so extreme was my
nervousness ; but when I once fairly began, my ideas
crowded upon me almost faster than I could write them.
And so we all sat, nothing heard but the scratching of
the pens and the occasional crackle of the examiner's
Times as he quietly looked over the news of the day.

The examination began at eleven. At two they
brought in lunch. It was a good meal enough, but the
circumstances were not particularly favourable to enjoy-
ment, so after a short delay we resumed our work. It
began to be evident between whom the contest lay, and
the others determined that I was one man's competitor
and Stocks [2] (he is now in the East India service) the
other. Scratch, scratch, scratch ! Four o'clock came,
the usual hour of closing the examination, but Stocks

---

[1] His eldest sister Mrs. Scott.
[2] John Ellerton Stocks, M.D., London, distinguished himself
as a botanist in India. He travelled and collected in Beloochistan
and Scinde ; died 1854.

and I had not half done, so with the consent of the
others we petitioned for an extension. The examiner
was willing to let us go on as long as we liked. Never
did I see man write like Stocks ; one might have taken
him for an attorney's clerk writing for his dinner. We
went on. I had finished a little after eight, he went on
till near nine, and then we had tea and dispersed.

Great were the greetings I received when I got home,
where my long absence had caused some anxiety. The
decision would not take place for some weeks, and many
were the speculations made as to the probabilities of
success. I for my part managed to forget all about it,
and went on my ordinary avocations without troubling
myself more than I could possibly help about it. I knew
too well my own deficiencies to have been either surprised or disappointed at failure, and I made a point of
shattering all involuntary "castles in the air" as soon as
possible. My worst anticipations were realised. One
day S. came to me with a sorrowful expression of
countenance. He had inquired of the Beadle as to the
decision, and ascertained on the latter's authority that
all the successful candidates were University College
men, whereby, of course, I was excluded. I said, "Very
well, the thing was not to be helped," put my best face
upon the matter, and gave up all thoughts of it. Lizzie,
too, came to comfort me, and, I believe, felt it more
than I did. What was my surprise on returning home
one afternoon to find myself suddenly seized, and the
whole female household vehemently insisting on kissing
me. It appeared an official-looking letter had arrived
for me, and Lizzie, as I did not appear, could not restrain
herself from opening it. I was second, and was to
receive a medal [1] accordingly, and dine with the guild on
the 9th November to have it bestowed.

I dined with the company, and bore my share in

---

[1] Silver Medal of the Apothecaries' Society, 9th November

both pudding and praise, but the charm of success lay in Lizzie's warm congratulation and sympathy. Since then she always took upon herself to prophesy touching the future fortunes of "the boy."

The haphazard, unsystematic nature of preliminary medical study here presented cannot fail to strike one with wonder. Thomas Huxley was now seventeen; he had already had two years' "practice in pharmacy," as a testimonial put it. After a similar apprenticeship, his brother had made the acquaintance of the director of the Gloucester Lunatic Asylum, and was given by him the post of dispenser or "apothecary," which he filled so satisfactorily as to receive a promise that if he went to London for a couple of years to complete his medical training, a substitute should be appointed meanwhile to keep the place until he returned.

The opportunity to which both the brothers looked came in the shape of the Free Scholarships offered by the Charing Cross Hospital to students whose parents were unable to pay for their education. Testimonials as to the position and general education of the candidates were required, and it is curious that one of the persons applied to by the elder

1842. Another botanical prize is a book—*La Botanique*, by A. Richard—with the following inscription :—

THOMAE HUXLEY
In Exercitatione Botanicēs
Apud Scholam Collegii Sydenhamiensis
Optime Merenti
Hunc librum dono dedit
RICARDUS D. HOBLYN, Botanicēs Professor.

Huxley was J. H. Newman, at that time Vicar of Littlemore, who had been educated at Dr. Nicholas' School at Ealing; but he refused, having no personal acquaintance with the boys.

The application for admission to the lectures and other teaching at the Hospital states of the young T. H. Huxley that "He has a fair knowledge of Latin, reads French with facility, and knows something of German. He has also made considerable progress in the Mathematics, having, as far as he has advanced, a thorough not a superficial knowledge of the subject." The document ends in the following confident words :—

I appeal to the certificates and testimonials that will be herewith submitted for evidence of their past conduct, offering prospectively that these young men, if elected to the Free Scholarships of the Charing Cross Hospital and Medical College, will be diligent students, and in all things submit themselves to the controul and guidance of the Director and Medical Officers of the establishment. A father may be pardoned, perhaps, for adding his belief that these young men will hereafter reflect credit on any institution from which they may receive their education.

The authorities replied that "although it is not usual to receive two members of the same family at the same time, the officers, taking into consideration the age of Mr. Huxley, sen., the numerous and satisfactory testimonials of his respectability, and of the good conduct and merits of the candidates, have decided upon admitting Mr. J. E. and Mr. T. Huxley on this occasion."

The brothers began their hospital course on October 1, 1842. Here, after a time, my father seems to have begun working more steadily and systematically than he had done before, under the influence of a really good teacher.

Looking back (he says) on my "Lehrjahre," I am sorry to say that I do not think that any account of my doings as a student would tend to edification. In fact, I should distinctly warn ingenuous youth to avoid imitating my example. I worked extremely hard when it pleased me, and when it did not, which was a very frequent case, I was extremely idle (unless making caricatures of one's pastors and masters is to be called a branch of industry), or else wasted my energies in wrong directions. I read everything I could lay hands upon, including novels, and took up all sorts of pursuits to drop them again quite as speedily. No doubt it was very largely my own fault, but the only instruction from which I obtained the proper effect of education was that which I received from Mr. Wharton Jones, who was the lecturer on physiology at the Charing Cross School of Medicine. The extent and precision of his knowledge impressed me greatly, and the severe exactness of his method of lecturing was quite to my taste. I do not know that I have ever felt so much respect for anybody as a teacher before or since. I worked hard to obtain his approbation, and he was extremely kind and helpful to the youngster who, I am afraid, took up more of his time than he had any right to do. It was he who suggested the publication of my first scientific paper—a very little one—in the *Medical Gazette* of 1845, and most kindly corrected the literary faults which abounded in it, short as it was ; for at that time, and for many years afterwards, I detested the trouble of writing, and would take no pains with it.

He never forgot his debt to Wharton Jones, and

years afterwards was delighted at being able to do
him a good turn, by helping to obtain a pension for
him.   But although in retrospect he condemns the
fitfulness of his energies and his want of system,
which left much to be learned afterwards which
might with advantage have been learned then, still
it was his energy that struck his contemporaries.
I have a story from one of them that when the other
students used to go out into the court of the hospital
after lectures were over, they would invariably catch
sight of young Huxley's dark head at a certain
window bent over a microscope while they amused
themselves outside.   The constant silhouette framed
in the outlines of the window tickled the fancy of
the young fellows, and a wag amongst them dubbed
it with a name that stuck, "The Sign of the Head
and Microscope."

The scientific paper, too, which he mentions, was
somewhat remarkable under the circumstances.   It
is not given to every medical student to make an
anatomical discovery, even a small one.   In this case
the boy of nineteen, investigating things for himself,
found a hitherto undiscovered membrane in the root
of the human hair, which received the name of
Huxley's layer.

Speculations, too, such as had filled his mind in
early boyhood, still haunted his thoughts.   In one
of his letters from the *Rattlesnake*, he gives an account
of how he was possessed in his student days by that
problem which has beset so many a strong imagina-

tion, the problem of perpetual motion, and even
sought an interview with Faraday, whom he left
with the resolution to meet the great man some day
on a more equal footing.

March 1848.

To-day, ruminating over the manifold ins and outs of
life in general, and my own in particular, it came into
my head suddenly that I would write down my interview
with Faraday—how many years ago ? Aye, there's the
rub, for I have completely forgotten. However, it must
have been in either my first or second winter session at
Charing Cross, and it was before Christmas, I feel sure.

I remember how my long brooding perpetual motion
scheme (which I had made more than one attempt to
realise, but failed owing to insufficient mechanical dex-
terity) had been working upon me, depriving me of rest
even, and heating my brain with *châteaux d'Espagne* of
endless variety. I remember, too, it was Sunday morning
when I determined to put the questions, which neither
my wits nor my hands would set at rest, into some hands
for decision, and I determined to go before some tribunal
from whence appeal should be absurd.

But to whom to go ? I knew no one among the high
priests of science, and going about with a scheme for
perpetual motion was, I knew, for most people the same
thing as courting ridicule among high and low. After
all I fixed upon Faraday, possibly perhaps because I
knew where he was to be found, but in part also because
the cool logic of his works made me hope that my poor
scheme would be treated on some other principle than
that of mere previous opinion one way or other. Besides,
the known courtesy and affability of the man encouraged
me. So I wrote a letter, drew a plan, enclosed the two
in an envelope, and tremblingly betook myself on the
following afternoon to the Royal Institution.

" Is Dr. Faraday here ? " said I to the porter. " No,

sir, he has just gone out." I felt relieved. "Be good
enough to give him this letter," and I was hurrying out
when a little man in a brown coat came in at the glass
door. "Here is Dr. Faraday," said the man, and gave
him my letter. He turned to me and courteously inquired what I wished. "To submit to you that letter,
sir, if you are not occupied." "My time is always
occupied, sir, but step this way," and he led me into the
museum or library, for I forget which it was, only I
know there was a glass case against which we leant. He
read my letter, did not think my plan would answer.
Was I acquainted with mechanism, what we call the laws
of motion? I saw all was up with my poor scheme, so
after trying a little to explain, in the course of which I
certainly failed in giving him a clear idea of what I
would be at, I thanked him for his attention, and went
off as dissatisfied as ever. The sense of one part of the
conversation I well recollect. He said "that were the
perpetual motion possible, it would have occurred spon-
taneously in nature, and would have overpowered all
other forces," or words to that effect. I did not see the
force of this, but did not feel competent enough to discuss
the question.

However, all this exorcised my devil, and he has
rarely come to trouble me since. Some future day,
perhaps, I may be able to call Faraday's attention more
decidedly. Perge modo! "wie das Gestirn, ohne Hast,
ohne Rast" (Das Gestirn in a midshipman's berth!).

In other respects also his student's career was a
brilliant one. In 1843 he won the first chemical
prize, the certificate stating that his "extraordinary
diligence and success in the pursuit of this branch of
science do him infinite honour." At the same time,
he also won the first prize in the class of anatomy
and physiology. On the back of Wharton Jones'

certificate is scribbled in pencil: "Well, 'tis no matter. Honour pricks me on. Yea, but how if honour prick me off when I come on? How then?"

Finally, in 1845 he went up for his M.B. at London University, and won a gold medal for anatomy and physiology, being second in honours in that section.

Whatever then he might think of his own work, judged by his own standards, he had done well enough as medical students go. But a brilliant career as a student did not suffice to start him in life or provide him with a livelihood. How he came to enter the Navy is best told in his own words.

It was in the early spring of 1846, that, having finished my obligatory medical studies and passed the first M.B. examination at the London University, though I was still too young to qualify at the College of Surgeons, I was talking to a fellow-student (the present eminent physician, Sir Joseph Fayrer), and wondering what I should do to meet the imperative necessity for earning my own bread, when my friend suggested that I should write to Sir William Burnett, at that time Director-General for the Medical Service of the Navy, for an appointment. I thought this rather a strong thing to do, as Sir William was personally unknown to me, but my cheery friend would not listen to my scruples, so I went to my lodgings and wrote the best letter I could devise. A few days afterwards I received the usual official circular of acknowledgment, but at the bottom there was written an instruction to call at Somerset House on such a day. I thought that looked like business, so at the appointed time I called and sent in my card while I waited in Sir William's anteroom.

He was a tall, shrewd-looking old gentleman, with a broad Scotch accent, and I think I see him now as he entered with my card in his hand. The first thing he did was to return it, with the frugal reminder that I should probably find it useful on some other occasion. The second was to ask whether I was an Irishman. I suppose the air of modesty about my appeal must have struck him. I satisfied the Director-General that I was English to the backbone, and he made some inquiries as to my student career, finally desiring me to hold myself ready for examination. Having passed this, I was in Her Majesty's Service, and entered on the books of Nelson's old ship, the *Victory*, for duty at Haslar Hospital, about a couple of months after my application.

My official chief at Haslar was a very remarkable person, the late Sir John Richardson, an excellent naturalist and far-famed as an indomitable Arctic traveller. He was a silent, reserved man, outside the circle of his family and intimates ; and having a full share of youthful vanity, I was extremely disgusted to find that " Old John," as we irreverent youngsters called him, took not the slightest notice of my worshipful self, either the first time I attended him, as it was my duty to do, or for some weeks afterwards. I am afraid to think of the lengths to which my tongue may have run on the subject of the churlishness of the chief, who was, in truth, one of the kindest-hearted and most considerate of men. But one day, as I was crossing the hospital square, Sir John stopped me and heaped coals of fire on my head by telling me that he had tried to get me one of the resident appointments, much coveted by the assistant-surgeons, but that the Admiralty had put in another man. " However," said he, " I mean to keep you here till I can get you something you will like," and turned upon his heel without waiting for the thanks I stammered out. That explained how it was I had not been packed off to the West Coast of Africa like some of my juniors, and why, eventually, I remained altogether seven months at Haslar.

After a long interval, during which "Old John" ignored my existence almost as completely as before, he stopped me again as we met in a casual way, and describing the service on which the *Rattlesnake* was likely to be employed, said that Captain Owen Stanley, who was to command the ship, had asked him to recommend an assistant-surgeon who knew something of science ; would I like that ? Of course I jumped at the offer. "Very well, I give you leave ; go to London at once and see Captain Stanley." I went, saw my future commander, who was very civil to me, and promised to ask that I should be appointed to his ship, as in due time I was. It is a singular thing that during the few months of my stay at Haslar I had among my messmates two future Directors-General of the Medical Service of the Navy (Sir Alexander Armstrong and Sir John Watt-Reid), with the present President of the College of Physicians, and my kindest of doctors, Sir Andrew Clark.

A letter to his eldest sister, Lizzie, dated from Haslar, May 24, 1846, shows how he regarded the prospect now opening before him.

. . . As I see no special queries in your letter, I think I shall go on to tell you what that same way of life is likely to be—my fortune having already been told for me (for the next five years at least). I told you in my last that I was likely to have a permanency here. Well, I was recommended by Sir John Richardson, and should have certainly had it, had not (luckily) the Admiralty put in a man of their own. Having a good impudent faith in my own star (Wie das Gestirn, ohne Hast, ohne Rast), I knew this was only because I was to have something better, and so it turned out ; for a day or two after I was ousted from the museum, Sir J. Richardson (who has shown himself for some reason or another a special good friend to me) told me that he had received a letter from Captain Owen Stanley, who is to command an *exploring expedition* to New Guinea (not coast of Africa, mind), requesting him to recommend an assistant surgeon for this expedition—would I like the appointment? As you may imagine, I was delighted at the offer, and immediately accepted it. I was recommended accordingly to Captain Stanley and Sir W. Burnett, and I shall be appointed as soon as the ship is in commission. We are to have the *Rattlesnake*, a 28-gun frigate, and as she will fit out here, I shall have no trouble. We sail probably in September.

New Guinea, as you may be aware, is a place almost unknown, and our object is to bring back a full account of its Geography, Geology, and Natural History. In the latter department with which I shall have (in addition to my medical functions) somewhat to do, we shall form one grand collection of specimens and deposit it in the British Museum or some other public place, and this main object being always kept in view, we are at liberty to collect and work for ourselves as we please. Depend upon it, unless some sudden attack of laziness supervenes, such an opportunity shall not slip unused out of my hands. The great difficulty in such a wide field is to choose an object. In this point, however, I hope to be greatly assisted by the scientific folks, to many of whom I have already had introductions (Owen, Gray, Grant, Forbes), and this, I assure you, I look upon as by no means the least of the advantages I shall derive from being connected with the expedition. I have been twice to town to see Captain Stanley. He is a son of the Bishop of Norwich, is an exceedingly gentlemanly man, a thorough scientific enthusiast, and shows himself altogether very much disposed to forward my views in every possible way. Being a scientific man himself, he will take care to have the ship's arrangements as far as possible in harmony with scientific pursuits—a circumstance you would appreciate as highly as I do if you were as well acquainted as I now am with the ordinary opportunities of an assistant surgeon.   Furthermore, I am given to understand that if one does anything at all, promotion is almost certain. So that altogether I am in a very fair way, and would snap my fingers at the Grand Turk.   Wharton Jones was delighted when I told him about my appointment.   Dim visions of strangely formed corpuscles seemed to cross his imagination like the ghosts of the kings in *Macbeth.*

> What seems his head
> The likeness of a nucleated cell has on.

The law's delays are proverbial, but on this occasion, as on the return of the *Rattlesnake,* the Admiralty seem to have been almost as provoking to the eager young surgeon as any lawyer could have been.   The appointment was promised in May; it was not made till October.   On the 6th of that month there is another letter to his sister, giving fuller particulars of his prospects on the voyage :—

My dearest Lizzie—At last I have really got my appointment and joined my ship.   I was so completely disgusted with the many delays that had occurred, that I made up my mind not to write to anybody again until I had my commission in my hand.   Henceforward, like another Jonah, my dwelling-place will be the " inwards " of the *Rattlesnake,* and upon the whole I really doubt whether Jonah was much worse accommodated, so far as room goes, than myself.   My total length, as you are aware, is considerable, 5 feet 11 inches, possibly, but the height of the lower deck of the *Rattlesnake,* which will be my especial location, is at the outside 4 feet 10 inches. What I am to do with the superfluous foot I cannot divine. Happily, however, there is a sort of skylight into the berth, so that I shall be able to sit with the body in it and my head out.

Apart from joking, however, this is not such a great matter, and it is the only thing I would see altered in the whole affair. The officers, as far as I have seen them, are a very gentlemanly, excellent set of men, and considering we are to be together for four or five years, that is a matter of no small importance. I am not given to be sanguine, but I confess I expect a good deal to arise out of this appointment. In the first place, surveying ships are totally different from the ordinary run of men-of-war. The requisite discipline is kept up, but not in the martinet style. Less form is observed. From the men who are appointed having more or less scientific turns, they have more respect for one another than that given by mere position in the service, and hence that position is less taken advantage of. They are brought more into contact, and hence those engaged in the surveying service almost proverbially stick by one another. To me, whose interest in the service is almost all to be made, this is a matter of no small importance.

Then again, in a surveying ship you can work. In an ordinary frigate, if a fellow has the talents of all the scientific men from Archimedes downwards compressed into his own peculiar skull, they are all lost. Even if it were possible to study in a midshipman's berth, you have not room in your "chat" for more than a dozen books. But in the *Rattlesnake* the whole poop is to be converted into a large chart-room with bookshelves and tables and plenty of light. There I may read, draw, or microscopise at pleasure, and as to books, I have a *carte blanche* from the Captain to take as many as I please, of which permission we shall avail ourself—rather—and besides all this, from the peculiar way in which I obtained this appointment, I shall have a much wider swing than assistant surgeons in general get. I can see clearly that certain

branches of the natural history work will fall into my hands
if I manage properly through Sir John Richardson, who has
shown himself a very kind friend all throughout, and also
through Captain Stanley I have been introduced to several
eminent zoologists—to Owen and Gray and Forbes of
King's College.   From all these men much is to be learnt
which becomes peculiarly my own, and can of course
only be used and applied by me.   From Forbes especially
I have learned and shall learn much with respect to
dredging operations (which bear on many of the most
interesting points of zoology).   In consequence of this I
may very likely be entrusted with the carrying of them
out, and all that is so much the more towards my opportunities.   Again, I have learnt the calotype process for
the express purpose of managing the calotype apparatus,
for which Captain Stanley has applied to the Government.

And having once for all enumerated all these meaner
prospects of mere personal advancement, I must confess I
do glory in the prospect of being able to give myself up
to my own favourite pursuits without thereby neglecting
the proper duties of life.   And then perhaps by the
following of my favourite motto—

> Wie das Gestirn,
> Ohne Hast,
> Ohne Rast—

something may be done, and some of Sister Lizzie's fond
imaginations turn out not altogether untrue.

I perceive that I have nearly finished a dreadfully
egotistical letter, but I know you like to hear of my
doings, so shall not apologise.   Kind regards to the
Doctor and kisses to the babbies.   Write me a long letter
all about yourselves.—Your affect. brother,

T. H. HUXLEY.

One more description to complete the sketch of
his quarters on board the *Rattlesnake*.   It is from a

letter to his mother, written at Plymouth, where the *Rattlesnake* put in after leaving Portsmouth. The comparison with the ordinary quarters of an assistant-surgeon, and the shifts to which a studious man might be put in his endeavour to find a quiet spot to work in, have a flavour of Mr. Midshipman Easy about them to relieve the deplorable reality of his situation :—

You will be very glad to know that I am exceedingly comfortable here. My cabin has now got into tolerable order, and what with my books—which are, I am happy to say, not a few—my gay curtain and the spicy oilcloth which will be down on the floor, looks most respectable. Furthermore, although it is an unquestionably dull day, I have sufficient light to write here, without the least trouble, to read, or even if necessary, to use my microscope. I went to see a friend of mine on board the *Recruit* the other day, and truly I hugged myself when I compared my position with his. The berth where he and seven others eat their daily bread is hardly bigger than my cabin, except in height—and, of course, he has to sleep in a hammock. My friend is rather an eccentric character, and, being missed in the ship, was discovered the other day reading in the main-top—the only place, as he said, sufficiently retired for study. And this is really no ex-aggeration. If I had no cabin, I should take to drinking in a month.

It was during this period of waiting that he attended his first meeting of the British Association, which was held in 1846 at Southampton. Here he obtained from Professor Edward Forbes one of his living specimens of *Amphioxus lanceolatus,* and made an examination of its blood.   The result was a short paper read at the following meeting of the Association,[1] which showed that in the composition of its blood this lowly vertebrate approached very near the invertebrates.

[1] "Examination of the Corpuscles of the Blood of Amphioxus lanceolatus," *British Association Report*, 1847, ii. p. 95, and *Sci. Memoirs*, i.

CHAPTER III

1846–1849

IT is a curious coincidence that, like two other leaders of science, Charles Darwin and Joseph Dalton Hooker, their close friend Huxley began his scientific career on board one of Her Majesty's ships. He was, however, to learn how little the British Government of that day, for all its professions, really cared for the advancement of knowledge.[1] But of the immense value to himself of these years of hard training, the discipline, the knowledge of men and of the capabilities of life, even without more than the barest

[1] The key to this attitude on the part of the Admiralty is to be found in the scathing description in Briggs' *Naval Administration from 1827 to 1892*, p. 92, of the ruinous parsimony of either political party at this time with regard to the navy—a policy the results of which were only too apparent at the outbreak of the Crimean War. I quote a couple of sentences, "The navy estimates were framed upon the lowest scale, and reduction pushed to the very verge of danger." "Even from a financial point of view the course pursued was the reverse of economical, and ultimately led to wasteful and increased expenditure." Thus the liberal professions of the Admiralty were not fulfilled ; its goodwill gave the young surgeon three and a half years of leave from active service ; with an obdurate Treasury, it could do no more.

necessities of existence—of this he often spoke. As he puts it in his Autobiography :—

Life on board Her Majesty's ships in those days was a very different affair from what it is now, and ours was exceptionally rough, as we were often many months without receiving letters or seeing any civilised people but ourselves. In exchange, we had the interest of being about the last voyagers, I suppose, to whom it could be possible to meet with people who knew nothing of fire-arms—as we did on the south coast of New Guinea—and of making acquaintance with a variety of interesting savage and semi-civilised people. But, apart from experience of this kind and the opportunities offered for scientific work, to me, personally, the cruise was extremely valuable. It was good for me to live under sharp discipline ; to be down on the realities of existence by living on bare necessaries : to find how extremely well worth living life seemed to be when one woke up from a night's rest on a soft plank, with the sky for canopy, and cocoa and weevilly biscuit the sole prospect for breakfast ; and, more especially, to learn to work for the sake of what I got for myself out of it, even if it all went to the bottom and I along with it. My brother officers were as good fellows as sailors ought to be and generally are, but, naturally, they neither knew nor cared anything about my pursuits, nor understood why I should be so zealous in pursuit of the objects which my friends, the middies, christened "Buffons," after the title conspicuous on a volume of the *Suites à Buffon*, which stood on my shelf in the chart-room.

On the whole, life among the company of officers was satisfactory enough.[1] Huxley's immediate

[1] The Assistant-Surgeon messed in the gun-room with the middies. A man in the midst of a lot of boys, with hardly any grown-up companions, often has a rather unenviable position ; but

superior, John Thomson, was a man of sterling worth ; and Captain Stanley was an excellent commander, and sympathetic withal. Among Huxley's messmates there was only one, the ship's clerk, who ever made himself actively disagreeable, and a quarrel with him only served to bring into relief the young surgeon's integrity and directness of action. After some dispute, in which he had been worsted, this gentleman sought to avenge himself by dropping mysterious hints as to Huxley's conduct before joining the ship. He had been treasurer of his mess ; there had been trouble about the accounts, and a scandal had barely been averted. This was not long in coming to Huxley's ears. Furiously indignant as he was, he did not lose his self-control ; but promptly inviting the members of the wardroom to meet as a court of honour, laid his case before them, and challenged his accuser to bring forward any tittle of evidence in support of his insinuations. The latter had nothing to say for himself, and made a formal retraction and apology. A signed account of the proceedings was kept by the first officer, and a duplicate by Huxley, as a defence against any possible revival of the slander.

On December 3, 1846, the *Rattlesnake* frigate left Spithead, but touched again at Plymouth to ship £65,000 of specie for the Cape. This delay was no

says Captain Heath, who was one of these middies, Huxley's constant good spirits and fun, when he was not absorbed in his work, his freedom from any assumption of superiority over them, made the boys his good comrades and allies.

pleasure to the young Huxley; it only served to renew the pain of parting from home, so that, after writing a last letter to reassure his mother as to the comfort of his present quarters, he was glad to lose sight of the English coast on the 11th.

Madeira was reached on the 18th.   On the 26th they sailed for Rio de Janeiro, where they stayed from January 23 to February 2, 1847.   Here Huxley had his first experience of tropical dredging in Botafago Bay, with Macgillivray, naturalist to the expedition. It was a memorable occasion, the more so, because in the absence of a sieve they were compelled to use their hands as strainers the first day.   Happily the want was afterwards supplied by a meat cover.   From the following letter it seems that several prizes of value were taken in the dredge :—

RIO JANEIRO, *Jan.* 24, 1847.

MY DEAR MOTHER—Four weeks of lovely weather and uninterrupted fair winds brought us to this southern fairyland.   In my last letter I told you a considerable yarn about Madeira, I guess, and so for fear lest you should imagine me scenery mad, I will spare you any description of Rio Harbour.   Suffice it to say that it contends with the Bay of Naples for the title of the most beautiful place in the world.   It must beat Naples in luxuriance and variety of vegetation, but from all accounts, to say nothing of George's [1] picture, falls behind it in the colours of sky and sea, that of the latter being in the harbour and for some distance outside of a dirty olive green like the washings of a painter's palette.

We have come in for the purpose of effecting some

---

[1] His eldest brother.

trifling repairs, which, though not essential to the safety
of the ship, will nevertheless naturally enhance the
comfort of its inmates. This you will understand when
I tell you that in consequence of these same defects I
have had water an inch or two deep in my cabin, wish-
washing about ever since we left Madeira.

We crossed the line on the 13th of this month, and as
one of the uninitiated I went through the usual tomfoolery
practised on that occasion. The affair has been too often
described for me to say anything about it. I had the
good luck to be ducked and shaved early, and of course
took particular care to do my best in serving out the unhappy beggars who had to follow. I enjoyed the fun
well enough at the time, but unquestionably it is on all
grounds a most pernicious custom. It swelled our sick
list to double the usual amount, and one poor fellow, I
am sorry to say, died of the effects of pleurisy then
contracted.

We have been quite long enough at sea now to enable
me to judge how I shall get on in the ship, and to form
a very clear idea of how it fits me and how I fit it. In
the first place, I am exceedingly well and exceedingly contented with my lot. My opinion of the advantages lying
open to me increases rather than otherwise as I see my way
about me. I am on capital terms with all the superior
officers, and I find them ready to give me all facilities. I
have a place for my books and microscope in the chart-
room, and there I sit and read in the morning much as
though I were in my rooms in Agar Street. My immediate superior, Johnny Thomson, is a long-headed
good fellow, without a morsel of humbug about him—a
man whom I thoroughly respect, both morally and intellectually. I think it will be my fault if we are not
fast friends through the commission. One friend on
board a ship is as much as anybody has a right to expect.

It is just the interval between the sea and the land
breezes, the sea like glass, and not a breath stirring. I

shall become soup if I do not go on deck. Temp. at noon 86 in shade, 139 in sun. *N.B.*—It has been up to 89 in shade, 139 in sun since this.

March 28.—I see I concluded with a statement of temp. Since then it has been considerably better—140 in sun ; however, in the shade it rarely rises above 86 or so, and when the sea or land breezes are blowing this is rather pleasant than otherwise.

I have been ashore two or three times. The town is like most Portuguese towns, hot and stinking, the odours here being improved by a strong flavour of nigger from the slaves, of whom there is an immense number. They seem to do all the work, and their black skins shine in the sun as though they had been touched up with Warren, 30 Strand. They are mostly in capital condition, and on the whole look happier than the corresponding class in England, the manufacturing and agricultural poor, I mean. I have a much greater respect for them than for their beastly Portuguese masters, than whom there is not a more vile, ignorant, and besotted nation under the sun. I only regret that such a glorious country as this should be in such hands. Had Brazil been colonised by Englishmen, it would by this time have rivalled our Indian Empire.

The naturalist Macgillivray and I have had several excursions under pretence of catching butterflies, etc. On the whole, however, I think we have been most successful in imbibing sherry cobbler, which you get here in great perfection. By the way, tell Cooke,[1] with my kindest regards, that ——— is a lying old thief, many of the things he told me about Macgillivray, *e.g.*, being an ignoramus in natural history, etc. etc., having proved to be lies. He is at any rate a very good ornithologist, and, I can testify, is exceedingly zealous in his vocation as a collector. As in these (points) Mr. ———'s statements

---

[1] His brother-in-law.

are unquestionably false, I must confess I feel greatly inclined to disbelieve his other assertions.

March 29.—We sail hence on Sunday for the Cape, so I will finish up. If you have not already written to me at that place, direct your letters to H.M.S. *Rattlesnake*, Sydney (to wait arrival). We shall probably be at the Cape some weeks surveying, thence shall betake ourselves to the Mauritius, and leave a card on Paul and Virginia, thence on to Sydney ; but it is of no use to direct to any place but the last.

P.S.—The Rattlesnakes are not idle   We shall most likely have something to say to the English savans before long.   If I have any friz in the fire I will let you know.

He gives a fuller account of this forthcoming work in a letter to his sister, dated Sydney, August 1, 1847. The two papers in question, as appears from the briefest notice in the *Proceedings of the Linnean Society*, ascribing them to William (!) Huxley, were read in 1849 :—

In my last letter I think I mentioned to you that I had worked out and sent home to the President of the Linnæan Soc., through Capt. Stanley, an account of *Physalia*, or Portuguese man-of-war as it is called, an animal whose structure and affinities had never been properly worked out. The careful investigation I made gave rise to several new ideas covering the whole class of animals to which this creature belongs, and these ideas I have had the good fortune to have had many opportunities of working out in the course of our subsequent wanderings, so that I am provided with materials for a second paper far more considerable in extent, and embracing an altogether wider field. This second paper is now partly *in esse*—that is, written out—and partly *in posse*—that is, in my head ; but I shall send it before

leaving. Its title will be "Observations upon the
Anatomy of the Diphydæ, and upon the Unity of
Organisation of the Diphydæ and Physophoridæ," and it
will have lots of figures to illustrate it. Now when we
return from the north I hope to have collected materials
for a much bigger paper than either of these, and to
which they will serve as steps. If my present anticipations turn out correct, this paper will achieve one of the
great ends of Zoology and Anatomy, viz. the reduction of
two or three apparently widely separated and incongruous
groups into modifications of the single type, every step of
the reasoning being based upon anatomical facts. There!
Think yourself lucky you have only got that to read
instead of the slight abstract of all three papers with
which I had some intention of favouring you.[1]

But five years ago you threw a slipper after me for
luck on my first examination, and I must have you to
do it for everything else.

At the Cape a stay of a month was made, from
March 6 to April 10, and certain surveying work
was done, after which the *Rattlesnake* sailed for
Mauritius. In spite of the fact that the novelty of
tropical scenery had worn off, the place made a deep impression. He writes to his mother, May 15, 1847 :—

After a long and somewhat rough passage from the
Cape, we made the highland of the Isle of France on the
afternoon of the 3rd of this month, and passing round
the northern extremity of the island, were towed into
Port Louis by the handsomest of tugs about noon on the
4th. In my former letter I have spoken to you of the
beauty of the places we have visited, of the picturesque
ruggedness of Madeira, the fine luxuriance of Rio, and

---

[1] These papers are to be found in vol. i. of the *Scientific
Memoirs* of T. H. Huxley, p. 9.

the rude and simple grandeur of South Africa. Much of my admiration has doubtless arisen from the novelty of these tropical or semi-tropical scenes, and would be less vividly revived by a second visit. I have become in a manner *blasé* with fine sights and something of a critic. All this is to lead you to believe that I have really some grounds for the raptures I am going into presently about Mauritius. In truth it is a complete paradise, and if I had nothing better to do, I should pick up some pretty French Eve (and there are plenty) and turn Adam. *N.B.* There are *no* serpents in the island.

This island is, you know, the scene of St. Pierre's beautiful story of Paul and Virginia, over which I suppose most people have sentimentalised at one time or another of their lives. Until we reached here I did not know that the tale was like the lady's improver—a fiction founded on fact, and that Paul and Virginia were at one time flesh and blood, and that their veritable dust was buried at Pamplemousses in a spot considered as one of the lions of the place, and visited as classic ground. Now, though I never was greatly given to the tender and sentimental, and have not had any tendencies that way greatly increased by the elegancies and courtesies of a midshipman's berth,—not to say that, as far as I recollect, Mdlle. Virginia was a bit of a prude, and M. Paul a pump,—yet were it but for old acquaintance sake, I determined on making a pilgrimage. Pamplemousses is a small village about seven miles from Port Louis, and the road to it is lined by rows of tamarind trees, of cocoanut trees, and sugar-canes. I started early in the morning in order to avoid the great heat of the middle of the day, and having breakfasted at Port Louis, made an early couple of hours' walk of it, meeting on my way numbers of the coloured population hastening to market in all the varieties of their curious Hindoo costume. After some trouble I found my way to the "Tombeaux" as they call them. They are situated in a garden at the

back of a house now in the possession of one Mr. Geary,
an English mechanist, who puts up half the steam
engines for the sugar mills in the island. The garden is
now an utter wilderness, but still very beautiful; round
it runs a grassy path, and in the middle of the path on
each side towards the further extremity of the garden is
a funeral urn supported on a pedestal, and as dilapidated
as the rest of the affair. These dilapidations, as usual,
are the work of English visitors, relic-hunters, who are
as shameless here as elsewhere. I was exceedingly
pleased on the whole with my excursion, and when I
returned I made a drawing of the place, which I will
send some day or other.

Since this I have made, in company with our purser
and a passenger, Mr. King, a regular pedestrian trip to
see some very beautiful falls up the country.

Leaving Mauritius on May 17, they prolonged
their voyage to Sydney by being requisitioned to
take more specie to Hobart Town, so that Sydney
was not reached until July 16, eight months since
they had had news of home.

The three months spent in this first visit to
Sydney proved to be one of the most vital periods
in the young surgeon's career. From boyhood up,
vaguely conscious of unrest, of great powers within
him working to find expression, he had yet been to
a certain extent driven in upon himself. He had
been somewhat isolated from those of his own age
by his eagerness for problems about which they cared
nothing; and the tendency to solitude, the habit of
outward reserve imposed upon an unusually warm
nature, were intensified by the fact that he grew up
in surroundings not wholly congenial. One member

alone of his family felt with him that complete and
vivid sympathy which is so necessary to the full
development of such a nature. When he was fourteen this sister married and left home, but the bond
between them was not broken. In some ways it
was strengthened by the lad's love for her children;
by his grief, scarcely less than her own, at the death
of her eldest little girl. Moreover they were brought
into close companionship for a considerable time
when, after his dismal period of apprenticeship at
Rotherhithe—to which he could never look back
without a shudder—he came to work under her
husband. She had encouraged him in his studies;
had urged him to work for the Botanical prize at
Sydenham College; had brightened his life with her
sympathy, and believed firmly in the brilliant future
which awaited him—a belief which for her sake, if
for nothing else, he was eager to justify by his best
exertions.

He had not had, so far, much opportunity of
entering the social world; but his visit to Sydney
gave him an opportunity of entering a good society
to which his commission in the navy was a sufficient
introduction. He was eager to find friendships if he
could, for his reserve was anything but misanthropic.
It was not long before he made the acquaintance of
William Macleay, a naturalist of wide research and
great speculative ability;[1] and struck up a close

[1] William Sharp Macleay, 1792-1865, who went to Australia
in 1839, combined distinction in the diplomatic service with

friendship with William Fanning, one of the leading
merchants of the town, a friendship which was to have
momentous consequences.   For it was at Fanning's
house that he met his future wife, Miss Henrietta
Anne Heathorn, for whom he was to serve longer
and harder than Jacob thought to serve for Rachel,
but who was to be his help and stay for forty years,
in his struggles ready to counsel, in adversity to
comfort; the critic whose judgment he valued above
almost any, and whose praise he cared most to win;
his first care and his latest thought, the other self,
whose union with him was a supreme example of
mutual sincerity and devotion.

It was a case of love, if not actually at first sight,
yet of very rapid growth when he came to learn the
quiet strength and tenderness of her nature as displayed in the management of her sister's household.
A certain simplicity and directness united with an
unusual degree of cultivation, had attracted him from
the first.   Before coming to Australia, she had been
two years at school in Germany, and her knowledge
of German and of German literature brought them
together on common ground.   Things ran very
smoothly at the beginning, and the young couple,
whose united ages amounted to forty-four years,
became engaged.

The marriage was to take place on his promotion
to the rank of full surgeon—a promotion he hoped

scientific pursuits.   His fivefold or "circular" system was a
"forcedly artificial attempt at a system of natural classification."

to attain speedily at the conclusion of the voyage on the strength of his scientific work, for this was the inducement held out by the Admiralty to energetic subalterns. The following letter to his sister describes the situation :—

SYDNEY HARBOUR, *March* 21, 1848.

. . . I have deferred writing to you in the hope of knowing something from yourself of your doings and whereabouts, and now that we are on the eve of departing for a long cruise in Torres Straits, I will no longer postpone the giving you some account of "was ist geschehen" on this side of the world. We spent three months in Sydney, and a gay three months of it we had,—nothing but balls and parties the whole time. In this corner of the universe, where men of war are rather scarce, even the old *Rattlesnake* is rather a lion, and her officers are esteemed accordingly. Besides, to tell you the truth, we are rather agreeable people than otherwise, and can manage to get up a very decent turn-out on board on occasion. What think you of your grave, scientific brother turning out a ball-goer and doing the "light fantastic" to a great extent? It is a great fact, I assure you. But there is a method in my madness. I found it exceedingly disagreeable to come to a great place like Sydney and think there was not a soul who cared whether I was alive or dead; so I determined to go into what society was to be had and see if I could not pick up a friend or two among the multitude of the empty and frivolous. I am happy to say that I have had more success than I hoped for or deserved, and there are now two or three houses where I can go and feel myself at home at all times. But my "home" in Sydney is the house of my good friend Mr. Fanning, one of the first merchants in the place. But thereby hangs a tale which, of all people in the world, I must tell you. Mrs. Fanning has a sister, and the dear little sister and I

managed to fall in love with one another in the most
absurd manner after seeing one another—I will not tell
you how few times, lest you should laugh.   Do you
remember how you used to talk to me about choosing a
wife?   Well, I think that my choice would justify even
your fastidiousness. . . . I think you will understand
how happy her love ought to and does make me.   I fear
that in this respect indeed the advantage is on my side,
for my present wandering life and uncertain position
must necessarily give her many an anxious thought.
Our future is indeed none of the clearest.   Three years
at the very least must elapse before the *Rattlesnake*
returns to England, and then unless I can write myself
into my promotion or something else, we shall be just
where we were.   Nevertheless I have the strongest per-
suasion that four years hence I shall be married and
settled in England.   We shall see.

I am getting on capitally at present.   Habit, inclination, and now a sense of duty keep me at work, and the
nature of our cruise affords me opportunities such as
none but a blind man would fail to make use of.   I have
sent two or three papers home already to be published,
which I have great hopes will throw light upon some
hitherto obscure branches of natural history, and I have
just finished a more important one, which I intend to get
read at the Royal Society.   The other day I submitted
it to William Macleay (the celebrated propounder of the
Quinary system), who has a beautiful place near Sydney,
and, I hear, "werry much approves what I have done."
All this goes to the comforting side of the question, and
gives me hope of being able to follow out my favourite
pursuits in course of time, without hindrance to what is
now the main object of my life.   I tell Netty to look to
being a "Frau Professorin" one of these odd days, and
she has faith, as I believe she would have if I told her I
was going to be Prime Minister.

We go to the northward again about the 23rd of this

month (April), and shall be away for ten or twelve
months surveying in Torres Straits. I believe we are to
refit in Port Essington, and that will be the only place
approaching to civilisation that we shall see for the whole
of that time; and after July or August next, when a
provision ship is to come up to us, we shall not even get
letters. I hope and trust I shall hear from you before
then. Do not suppose that my new ties have made me
forgetful of old ones. On the other hand, these are if
anything strengthened. Does not my dearest Nettie love
you as I do! and do I not often wish that you could see
and love and esteem her as I know you would. We
often talk about you, and I tell her stories of old times.

Another letter, a year later, gives his mother the
answers to a string of questions which, mother-like,
she had asked him, thirsting for exact and minute
information about her future daughter-in-law :—

SYDNEY, *Feb.* 1, 1849.
(After describing how he had just come back from a
nine months' cruise)—First and foremost, my dear
mother, I must thank you for your very kind letter of
September 1848. I read the greater part of it to Nettie,
who was as much pleased as I with your kindly wishes
towards both of us. Now I suppose I must do my best
to answer your questions. First, as to age, Nettie is
about three months younger than myself—that is the
difference in *our years*, but she is *in fact* as much younger
than her years as I am older than mine. Next, as to
complexion she is exceedingly fair, with the Saxon
yellow hair and blue eyes. Then as to face, I really
don't know whether she is pretty or not. I have never
been able to decide the matter in my own mind. Sometimes I think she is, and sometimes I wonder how the
idea ever came into my head. Whether or not, her
personal appearance has nothing whatever to do with the

hold she has upon my mind, for I have seen hundreds of prettier women. But I never met with so sweet a temper, so self-sacrificing and affectionate a disposition, or so pure and womanly a mind, and from the perfectly intimate footing on which I stand with her family I have plenty of opportunities of judging. As I tell her, the only great folly I am aware of her being guilty of was the leaving her happiness in the hands of a man like myself, struggling upwards and certain of nothing.

As to my future intentions I can say very little about them. With my present income, of course, marriage is rather a bad look-out, but I do not think it would be at all fair towards N. herself to leave this country without giving her a wife's claim upon me. . . . It is very unlikely I shall ever remain in the colony. Nothing but a very favourable chance could induce me to do so.

Much must depend upon how things go in England. If my various papers meet with any success, I may perhaps be able to leave the service. At present, however, I have not heard a word of anything I have sent. Professor Forbes has, I believe, published some of Macgillivray's letters to him, but he has apparently forgotten to write to Macgillivray himself, or to me. So I shall certainly send him nothing more, especially as Mr. Macleay (of this place, and a great man in the naturalist world) has offered to get anything of mine sent to the Zoological Society.

In the paper mentioned in the letter of March 21, above ("On the Anatomy and Affinities of the Family of the Medusæ"), Huxley aimed at "giving broad and general views of the whole class, considered as organised upon a given type, and inquiring into its relations with other families," unlike previous observers whose patience and ability had been devoted rather to "stating matters of detail concerning par-

ticular genera and species." At the outset, section
8 (*Sci. Mem.* i. 11), he states—

> I would wish to lay particular stress upon the com-
> position of this (the stomach) and other organs of the
> Medusæ out of *two distinct membranes*, as I believe that it
> is one of the essential peculiarities of their structure, and
> that a knowledge of the fact is of great importance in
> investigating their homologies. I will call these two
> membranes as such, and independently of any modifica-
> tions into particular organs, "foundation membranes."

And in section 56 (p. 23) one of the general conclusions which he deduces from his observations, is

> That a Medusa consists essentially of two membranes
> inclosing a variously-shaped cavity, inasmuch as its
> various organs are so composed,

a peculiarity shared by certain other families of
zoophytes. This is the point which that eminent
authority, Professor G. J. Allman, had in his mind
when he wrote to call my attention

> to a fact which has been overlooked in all the notices I
> have seen, and which I regard as one of the greatest
> claims of his splendid work on the recognition of
> zoologists. I refer to his discovery that the body of the
> Medusæ is essentially composed of two membranes, an
> outer and an inner, and his recognition of these as the
> homologues of the two primary germinal leaflets in the
> vertebrate embryo. Now this discovery stands at the
> very basis of a philosophic zoology, and of a true concep-
> tion of the affinities of animals. It is the ground on
> which Haeckel has founded his famous Gastræa Theory,
> and without it Kowalesky could never have announced
> his great discovery of the affinity of the Ascidians and
> Vertebrates, by which zoologists had been startled.

CHAPTER IV

1848–1850

THE whole cruise of the *Rattlesnake* lasted almost precisely four years, her stay in Australian waters nearly three. Of this time altogether eleven months were spent at Sydney, namely, July 16 to October 11, 1847; January 14 to February 2, and March 9 to April 29, 1848; January 24 to May 8, 1849; and February 14 to May 2, 1850. The three months of the first northern cruise were spent in the survey of the Inshore Passage—the passage, that is, within the Great Barrier Reef for ships proceeding from India to Sydney. In 1848, while waiting for the right season to visit Torres Straits, a short cruise was made in February and March, to inspect the lighthouses in Bass' Straits. It was on this occasion that Huxley visited Melbourne, then an insignificant town, before the discovery of gold had brought a rush of immigrants.

The second northern cruise of 1848, which lasted nine months, had for its object the completion of the survey of the Inner Passage as far as New Guinea

and the adjoining archipelago. The third cruise in
1849-50 again lasted nine months, and continued the
survey in Torres Straits, the Louisiade archipelago,
and the south-eastern part of New Guinea. After
this the original plan was to make a fourth cruise,
filling up the charts of the Inner Passage on the east
coast, and surveying the straits of Alass between
Lombok and Sumbawa in the Malay Archipelago;
then, instead of returning to Sydney, to proceed to
Singapore and so home by the Cape. But these
plans were altered by the untimely death of Captain
Stanley on March 13, and the *Rattlesnake* sailed for
England direct in May 1850.

There was a great monotony about these cruises,
particularly to those who were not constantly engaged
in the active work of surveying. The ship sailed
slowly from place to place, hunting out reefs and
islets; a stay of a few days would be made at some
lonely island, while charting expeditions went out in
the boats or supplies of water and fresh fruits were
laid in. On the second expedition there were two
cases of scurvy on board by the time the mail from
Sydney reached the ship at Cape York with letters
and lime-juice, the first reminder of civilisation for
four months and a half. On this cruise there was
an unusual piece of interest in Kennedy's ill-fated
expedition, which the *Rattlesnake* landed in Rockingham Bay, and trusted to meet again at Cape York.
Happy it was for Huxley that his duties forbade him
to accept Kennedy's proposal to join the expedition.

After months of weary struggles in the dense scrub, Kennedy himself, who had pushed on for help with his faithful black man Jacky, was speared by the natives when almost in sight of Cape York; Jacky barely managed to make his way there through his enemies, and guided a party to the rescue of the two starved and exhausted survivors of the disease-stricken camp by the Sugarloaf Hill. It was barely time. Another hour, and they too would have been killed by the crowd of blackfellows who hovered about in hopes of booty, and were only dispersed for a moment by the rescue party.

On the third cruise there were a few adventures more directly touching the *Rattlesnake*. Twice the landing parties, including Huxley, were within an ace of coming to blows with the islanders of the Louisiades, and on one occasion a portly member of the gun-room, being cut off by these black gentry, only saved his life by parting with all his clothes as presents to them, and keeping them amused by an impromptu dance in a state of nature under the broiling sun, until a party came to his relief. At Cape York also, a white woman was rescued who had been made prisoner by the blacks from a wreck, and had lived among them for several years. Here, too, Huxley and Macgillivray made a trip inland, and were welcomed by a native chief, who saw in the former the returning spirit of his dead brother.

Throughout the voyage Huxley was busy with his pencil, and many lithographs from his drawings illustrate the account of the voyage afterwards published.   As to his scientific work, he was accumulating a large stock of observations, but felt rather sore about the papers which he had already sent home, for no word had reached him as to their fate, not even that they had been received or looked over by Forbes, to whom they had been consigned.   As a matter of fact, they had not been neglected, as he was to find out on his return; but meanwhile the state of affairs was not reassuring to a man whose dearest hopes were bound up in the reception he could win for these and similar researches.   Altogether, it was with no little joy that he turned his back on the sweltering heat of Torres Straits, on the great mountains of New Guinea, the Owen Stanley range, which had remained hidden from D'Urville in the *Astrolabe* to be discovered by the explorers on the *Rattlesnake*, and the far-stretching archipelago of the Louisiades, one tiny island in which still bears the name of Huxley, after the assistant-surgeon of the *Rattlesnake*.

A few extracts from letters of the time will give a more vivid idea of what the voyage was like.   The first is from a letter to his mother, dated February 1, 1849 :—

. . . I suppose you have wondered at the long intervals of my letters, but my silence has been forced. I wrote from Rockingham Bay in May, and from Cape York in October.   After leaving the latter place we have had no communication with any one but the folks at

Port Essington, which is a mere military post, without any certain means of communication with England. We were ten weeks on our passage from Port Essington to Sydney and touched nowhere, so that you may imagine we were pretty well tired of the sea by the time we reached Port Jackson.

Thank God we are now safely anchored in our old quarters, and for the next three months shall enjoy a few of those comforts that make life worth living. . . .

The only place we have visited since my last budget to you was Port Essington, a military post which has been an object of much attention for some time past in connection with the steam navigation between Sydney and India. It is about the most useless, miserable, ill-managed hole in Her Majesty's dominions. Placed fifteen miles inland on the swampy banks of an estuary out of reach of the sea breezes, it is the most insufferably hot and enervating place imaginable. The temperature of the water alongside the ship was from 88 to 90, *i.e.* about that of a moderately warm bath, so that you may fancy what it is on land. Added to this, the commandant is a litigious old fool, always at war with his officers, and endeavouring to make the place as much of a hell morally as it is physically. Little more than two years ago a detachment of sixty men came out to the settlement. At the parade on the Sunday I was there, there were just ten men present. The rest were invalided, dead, or sick. I have no hesitation in saying that half of this was the result of ill-management. The climate in itself is not particularly unhealthy. We were all glad to get away from the place.

Another is to his sister, under date Sydney, March 14, 1849 :—

By the way, I may as well give you a short account of our cruise. We started from here last May to survey

what is called the inner passage to India. You must
know that the east coast of Australia has running parallel
to it at distances of from five miles to seventy or eighty
an almost continuous line of coral reefs, the Great Barrier
as it is called. Outside this line is the great Pacific,
inside is a space varying in width as above, and cut up
by little islands and detached reefs. Now to get to
India from Sydney, ships must go either inside or outside
the Great Barrier. The inside passage has been called
the Inner Route in consequence of its desirability for
steamers, and our business has been to mark out this
Inner Route safely and clearly among the labyrinth-like
islands and reefs within the Barrier. And a parlous dull
business it was for those who, like myself, had no
necessary and constant occupation. Fancy for five
mortal months shifting from patch to patch of white sand
in latitude from 17 to 10 south, living on salt pork and
beef, and seeing no mortal face but our own sweet
countenances considerably obscured by the long beard
and moustaches with which, partly from laziness and
partly from comfort, we had become adorned. I
cultivated a peak in Charles I. style, which imparted a
remarkably peculiar and *triste* expression to my sunburnt
phiz, heightened by the fact that the aforesaid beard was,
I regret to say it, of a very questionable auburn—my
messmates called it red.

We convoyed a land expedition as far as the Rockingham
Bay in 17 south under a Mr. Kennedy, which was to
work its way up to Cape York in 11 south and there
meet us. A fine noble fellow poor Kennedy was too. I
was a good deal with him at Rockingham Bay, and
indeed accompanied him in the exploring trips which he
made for some four or five days in order to see how the
land lay about him. In fact we got on so well together
that he wanted me much to accompany him and join the
ship again at Cape York, and if the Service would have
permitted of my absence I should certainly have done so.

But it was well I did not. Out of thirteen men composing the party but three remain alive. The rest have perished by starvation or the spears of the natives. Poor Kennedy himself had, in company with the black fellow attached to the party, by dint of incredible exertions, pushed on until he came within sight of the provision vessel waiting his arrival at Cape York. But here, within grasp of his object, a large party of natives attacked and killed him. The black fellow alone reached Cape York with the news. The other two men who were saved were the sole survivors of the party Kennedy left behind him at a spot near the coast, and were picked up by the provision vessel when she returned.

You may be sure I am not sorry to return home. I say home advisedly, for my friend Fanning's house is as completely my home as it well can be. And then Nettie had not heard anything of me for six months, so that I have been petted and spoiled ever since we came in. . . . As I tell her, I fear she has rested her happiness on a very insecure foundation ; but she is full of hope and confidence, and to me her love is the faith that moveth mountains. We have, as you may be sure, a thousand difficulties in our way, but like Danton I take for my motto, " De l'audace et encore de l'audace et toujours de l'audace," and look forward to a happy termination, nothing doubting.

TO HIS MOTHER

(Announcing the probable time of his return.)

SYDNEY, *Feb.* 11, 1850.

I cannot at all realise the idea of our return. We have been leading such a semi-savage life for years past, such a wandering nomadic existence, that any other seems in a manner unnatural to me. Time was when I should have looked upon our return with unmixed joy; but so many new and strong ties have arisen to unite me with Sydney, that now when the anchor is getting up for England, I scarcely know whether to rejoice or to grieve. You must not be angry, my dear Mother; I have none the less affection for you or any other of those whom I love in England—only a very great deal for a certain little lassie whom I must leave behind me without clearly seeing when we are to meet again. You must remember the Scripture as my excuse, "A man shall leave his father and mother and cleave unto his" (I wish I could add) wife. Our long cruises are fine times for reflection, and during the last I determined that we would be terribly prudent and get married about 1870, or the Greek Kalends, or, what is about the same thing, whenever I am afflicted with the *malheur de richesse*.

People talk about the satisfaction of an approving conscience. Mine approves me intensely; but I'll be hanged if I see the satisfaction of it. I feel much more inclined to swear "worse than our armies in Flanders." . . . So far as my private doings are concerned, I hear very satisfactory news of them. I heard from an old messmate of mine at Haslar the other day that Dr. MacWilliam, F.R.S., one of our deputy-inspectors, had been talking about one of my papers, and gave him to understand that it was to be printed. Furthermore, he is a great advocate for the claims of assistant surgeons to ward-room rank, and all that sort of stuff, and, I am told, quoted me as an example! Henceforward I look upon the learned doctor as a man of sound sense and discrimination! Without joking, however, I am glad to have come under his notice, as he may be of essential use to me. I find myself getting horribly selfish, looking at everything with regard to the influence it may have on my grand objects.

Further descriptions of the voyage are to be drawn from an article in the *Westminster Review* for

January 1854 (vol. v.), in which, under the title of
"Science at Sea," Huxley reviewed the *Voyage of the
"Rattlesnake*" by Macgillivray, the naturalist to the
expedition, which had recently appeared. This book
gave very few descriptions of the incidents and life on
board, and so drew in many ways a colourless picture
of the expedition. This defect the reviewer sought
to remedy by giving extracts from the so-called "unpublished correspondence" of one of the officers—
sketches apparently written for the occasion—as well
as from an equally unpublished but more real journal
kept by the same hand.

The description of the ship herself, of her inadequate equipment for the special purposes she
was to carry out, of the officers' quiet contempt of
scientific pursuits, which not even the captain's
influence was able to subdue, of the illusory promises
of help and advancement held out by the Admiralty
to young investigators, makes a striking foil to the
spirit in which the Government of thirty years later
undertook a greater scientific expedition. Perhaps
some vivid recollections of this voyage did something
to better the conditions under which the later investigators worked.

Thus, p. 100:

In the year 1846, Captain Owen Stanley, a young and
zealous officer, of good report for his capabilities as a
scientific surveyor, was entrusted with the command of
the *Rattlesnake*, a vessel of six-and-twenty guns, strong
and seaworthy, but one of that class unenviably dis-

tinguished in the war-time as a "donkey-frigate." To
the laity it would seem that a ship journeying to unknown
regions, when the lives of a couple of hundred men may,
at any moment, depend upon her handiness in going
about, so as to avoid any suddenly discovered danger,
should possess the best possible sailing powers. The
Admiralty, however, makes its selection upon other
principles, and exploring vessels will be invariably found
to be the slowest, clumsiest, and in every respect the
most inconvenient ships which wear the pennant. In
accordance with the rule, such was the *Rattlesnake;*
and to carry out the spirit of the authorities more completely, she was turned out of Portsmouth dockyard in
such a disgraceful state of unfitness, that her lower deck
was continually under water during the voyage.

Again, p. 100 :

It is necessary to be provided with books of reference,
which are ruinously expensive to a private individual,
though a mere dewdrop in the general cost of the fitting-
out of a ship, especially as they might be kept in store,
and returned at the end of a commission, like other
stores. A hundred pounds would have well supplied the
Rattlesnake; but she sailed without a volume, an application made by her captain not having been attended to.

P. 103 :

Of all those who were actively engaged upon the
survey, the young commander alone was destined by inevitable fate to be robbed of his just reward. Care and
anxiety, from the mobility of his temperament, sat not so
lightly upon him as they might have done, and this,
joined to the physical debility produced by the enervating
climate of New Guinea, fairly wore him out, making him
prematurely old before much more than half of the
allotted span was completed. But he died in harness, the
end attained, the work that lay before him honourably

done. Which of us may dare to ask for more? He has raised an enduring monument in his works, and his epitaph shall be the grateful thanks of many a mariner threading his way among the mazes of the Coral Sea.

P. 104:

The world enclosed within the timbers of a man-of-war is a most remarkable community, hardly to be rendered vividly intelligible to the mere landsman in these days of constitutional government and freedom of the press.

Then follows a vigorous sketch of sea life from Chamisso, suggesting that the type of one's relation to the captain is to be found in Jean Paul's *Biography of the Twins*, who were united back to back. This sketch Huxley enforces by a passage from the imaginary journal aforesaid, "indited apparently when the chains were yet new and somewhat galled the writer," to judge from which "little alteration would seem to have taken place in nautical life" since Chamisso's voyage, thirty years before.

You tell me (he writes) that you sigh for my life of freedom and adventure; and that, compared with mine, the conventional monotony of your own stinks in your nostrils. My dear fellow, be patient, and listen to what I have to say; you will then, perhaps, be a little more content with your lot in life, and a little less desirous of mine. Of all extant lives, that on board a ship-of-war is the most artificial—whether necessarily so or not is a question I will not undertake to decide; but the fact is indubitable.

How utterly disgusted you get with one another! Little peculiarities which would give a certain charm and variety to social intercourse under any other circum-

stances, become sources of absolute pain, and almost un-
controllable irritation, when you are shut up with them
day and night. One good friend and messmate of mine
has a peculiar laugh, whose iteration on our last cruise
nearly drove me insane.

There is no being alone in a ship. Sailors are
essentially gregarious animals, and don't at all understand
the necessity under which many people labour—I among
the rest—of having a little solitary converse with one's
self occasionally.

Then, to a landsman fresh from ordinary society and
its peculiarly undemonstrative ways, there is something
very wonderful about naval discipline. I do not mean to
say that the subordination kept up is more than is
necessary, nor perhaps is it in reality greater than is to
be found in a college, or a regiment, or a large mercantile
house ; but it is made so *very* obvious. You not only feel
the bit, but you see it ; and your bridle is hung with
bells to tell you of its presence.

Your captain is a very different person, in relation to
his officers, from the colonel of a regiment ; he is a demi-
god, a Dalai lama, living in solitary state ; sublime, un-
approachable ; and the radiation of his dignity stretches
through all the other members of the nautical hierarchy ;
hence all sorts of petty intrigues, disputes, grumblings,
and jealousies, which, to the irreverent eye of an "idler,"
give to the whole little society the aspect of nothing so
much as the court of Prinz Irenæus in Kater Murr's
inestimable autobiography.

P. 107 *sq.* :

After describing the illusory promises of the
Admiralty and their grudging spirit towards the
scientific members of the expedition, he continues :—

These are the *facilities and encouragement* to science
afforded by the Admiralty ; and it cannot be wondered

at if the same spirit runs through its subordinate officers.

Not that there is any active opposition—quite the reverse. But it is a curious fact, that if you want a boat for dredging, ten chances to one they are always actually or potentially otherwise disposed of ; if you leave your towing-net trailing astern in search of new creatures, in some promising patch of discoloured water, it is, in all probability, found to have a wonderful effect in stopping the ship's way, and is hauled in as soon as your back is turned ; or a careful dissection waiting to be drawn may find its way overboard as a "mess."

The singular disrespect with which the majority of naval officers regard everything that lies beyond the sphere of routine, tends to produce a tone of feeling very unfavourable to scientific exertions. How can it be otherwise, in fact, with men who, from the age of thirteen, meet with no influence but that which teaches them that the "Queen's regulations and instructions" are the law and the prophets, and something more ?

It may be said, without fear of contradiction, that in time of peace the only vessels which are engaged in services involving any real hardship or danger are those employed upon the various surveys ; and yet the men of easy routine—harbour heroes—the officers of *regular* men-of-war, as they delight to be called, pretend to think surveying a kind of shirking—in sea-phrase, "sloping." It is to be regretted that the officers of the surveying vessels themselves are too often imbued with the same spirit ; and though, for shame's sake, they can but stand up for hydrography, they are too apt to think an alliance with other branches of science as beneath the dignity of their divinity—the "Service."

P. 112:

Any adventures ashore were mere oases, separated by whole deserts of the most wearisome *ennui*. For weeks, perhaps, those who were not fortunate enough to be living hard and getting fatigued every day in the boats were yawning away their existence in an atmosphere only comparable to that of an orchid-house, a life in view of which that of Mariana in the moated grange has its attractions.

For instance, consider this extract from the journal of one of the officers, date August 1849 :—

"Rain! rain! *encore et toujours*—I wonder if it is possible for the mind of man to conceive anything more degradingly offensive than the condition of us 150 men, shut up in this wooden box, and being watered with hot water, as we are now. It is no exaggeration to say *hot*, for the temperature is that at which people at home commonly take a hot bath. It rains so hard that we have caught seven tons of water in one day, and it is therefore impossible to go on deck, though, if one did, one's condition would not be much improved. A *hot* Scotch mist covers the sea and hides the land, so that no surveying can be done; moving about in the slightest degree causes a flood of perspiration to pour out; all energy is completely gone, and if I could help it I would not think even; it's too hot. The rain awnings are spread, and we can have no wind sails up; if we could, there is not a breath of wind to fill them; and consequently the lower and main decks are utterly unventilated: a sort of solution of man in steam fills them from end to end, and surrounds the lights with a lurid halo. It's too hot to sleep, and my sole amusement consists in watching the cockroaches, which are in a state of intense excitement and happiness. They manifest these feelings in a very remarkable manner—a sudden unanimous impulse seems to seize the obscene thousands which usually lurk hidden in the corners of my cabin. Out they rush, helter-skelter, and run over me, my table, and my desk; others, more vigorous, fly, quite regardless of consequences, until they hit against something, upon

which, half spreading their wings, they make their heads
a pivot and spin round in a circle, in a manner which
indicates a temporary aberration of the cockroach mind.
It is these outbreaks alone which rouse us from our
lassitude. Knocks are heard resounding on all sides, and
each inhabitant of a cabin, armed with a slipper, is seen
taking ample revenge upon the disturbers of his rest and
the destroyers of his body and clothes."

Here, on the other hand, is an oasis, a bartering scene
at Bruny Island, in the Louisiade :—

"We landed at the same place as before, and this
time the natives ran down prancing and gesticulating.
Many of them had garlands of green leaves round their
heads, knees, and ankles ; some wore long streamers
depending from their arms and ears and floating in the
wind as they galloped along, shaking their spears and
prancing just as boys do when playing at horses. They
soon surrounded us, shouting 'Kelumai! Kelumai!'
(their word for iron), and offering us all sorts of things
in exchange. One very fine athletic man, 'Kai-oo-why-
who-at' by name, was perfectly mad to get an axe, and
very soon comprehended the arrangements that were
made. Mr. Brady drew ten lines on the sand and laid
an axe down by them, giving K—— (I really can't write
that long name all over again) to understand by signs
that when there was a 'bahar' (yam) on every mark he
should have the axe. He comprehended directly, and
bolted off as fast as he could run, soon returning with his
hands full of yams, which he deposited one by one on
the appropriate lines ; then fearful lest some of the others
should do him out of the axe, he caught hold of Brady
by the arm, and would not let him go until yams enough
had been brought by the others to make up the number,
and the axe was handed over to him.

"Then was there a yell of delight! He jumped up
with the axe, flourished it, passed it to his companions,
tumbled down and rolled over, kicking up his heels in

the air, and finally, catching hold of me, we had a grand
waltz, with various *poses plastiques,* for about a quarter of
a mile. I daresay he was unsophisticated enough to
imagine that I was filled with sympathetic joy, but I
grieve to say that I was taking care all the while to
direct his steps towards the village, which, as we had as
yet examined none of their houses, I was most desirous
of entering under my friend's sanction. I think he
suspected something, for he looked at me rather dubiously
when I directed our steps towards the entrance in the
bush which led to the houses, and wanted me to go back ;
but I was urgent, so he gave way, and we both entered
the open space, where we were joined by two or three
others, and sat down under a cocoanut tree.

"I persuaded him to sit for his portrait (taking care
first that my back was against the tree and my pistols
handy), and we ate green cocoanuts together, at last
attaining to so great a pitch of intimacy that he made
me change names with him, calling himself 'Tamoo' (my
Cape York name), and giving me to understand that I
was to take his own lengthy appellation. When I did so,
and talked to him as 'Tamoo,' nothing could exceed the
delight of all around ; they patted me as you would a
child, and evidently said to one another, 'This really
seems to be a very intelligent white fellow.'

"Like the Cape York natives, they were immensely
curious to look at one's legs, asking permission, very
gently but very pressingly, to pull up the trouser, span-
ning the calf with their hands, drawing in their breath
and making big eyes all the while. Once, when the front
of my shirt blew open, and they saw the white skin of
my chest, they set up a universal shout. I imagine that
as they paint *their* faces black, they fancied that we in-
geniously coloured ours white, and were astonished to see
that we were really of that (to them) disgusting tint all over."

On May 2, 1850, the *Rattlesnake* sailed for the last

time out of Sydney harbour, bound for England by way of the Horn. In spite of his cheerful anticipations, Huxley was not to see his future wife again for five years more, when he was at length in a position to bid her come and join him. During the three years of their engagement in Australia, they had at least been able to see each other at intervals, and to be together for months at a time. In the long periods of absence, also, they had invented a device to cheat the sense of separation. Each kept a particular journal, to be exchanged when they met again, and only to be read, day by day, during the next voyage. But now it was very different, their only means of communication being the slow agency of the post, beset with endless possibilities of misunderstanding when it brought belated answers to questions already months old and out of date in the changed aspect of circumstances. These perils, however, they weathered, and it proves how deep in the moral nature of each the bond between them was rooted, that in the end they passed safely through the still greater danger of imperceptibly growing estranged from one another under the influence of such utterly different surroundings.

A kindly storm which forced the old ship to put into the Bay of Islands to repair a number of small leaks that rendered the lower deck uninhabitable, made it possible for Huxley to send back a letter that should reach Australia in one month instead of ten after his departure.

He utilised a week's stay here characteristically enough in an expedition to Waimate, the chief missionary station and the school of the native institutions (a sort of Normal School for native teachers), in order to judge of his own inspection what missionary life was like.

I have been greatly surprised in these good people (he writes). I had expected a good deal of *straight-hairedness* (if you understand the phrase) and methodistical puritanism, but I find it quite otherwise. Both Mr. and Mrs. Burrows seem very quiet and unpretending—straightforward folks desirous of doing their best for the people among whom they are placed.

One touch must not be allowed to pass unnoticed in his appreciation of the missionaries' unstudied welcome to the belated travellers, whose proper host was unable to take them in :—"tea unlimited and a blazing fire, *together with a very nice cat.*"

By July 12, midwinter of course in the southern hemisphere, they had rounded the Horn, and Huxley writes from that most desolate of British possessions, the Falkland Islands :—

I have great hopes of being able to send a letter to you, *via* California, even from this remote corner of the world. It is the Ultima Thule and no mistake. Fancy two good-sized islands with undulated surface and sometimes elevated hills, but without tree or bush as tall as a man. When we arrived the 8th inst. the barren uniformity was rendered still more obvious by the deep coating of snow which enveloped everything. How can I describe to you "Stanley," the sole town, metropolis, and seat of government ? It consists of a lot of black, low, weather-board houses scattered along the hillsides which rise round the harbour. One barnlike place is Government House, another the pensioners' barracks, rendered imposing by four field-pieces in front; others smaller are the residences of the colonel, surgeon, etc. In one particularly black and unpromising-looking house lives a Mrs. Sulivan, the wife of Captain Sulivan,[1] who surveyed these islands, and has settled out here. I asked myself if I could have had the heart to bring you to such a desolate place, and myself said "No." However, I believe she is very happy with her children. Sulivan is a fine energetic man, so I suppose if she loves him, well and good, and fancies (is she not a silly woman?) that she has her reward. Mrs. Stanley has gone to stay with them while the ship remains here, and I think I shall go and look them up under pretence of making a call. They say that the present winter is far more savage than the generality of Falkland Island winters, and it had need be, for I never felt anything so bitterly cold in my life. The thermometer has been down below 22, and shallow parts of the harbour even have frozen. Nothing to be done ashore. My rifle lies idle in its case; no chance of a shot at a bull, and one has to go away 20 miles to get hold even of the upland geese and rabbits. The only thing to be done is to eat, eat, eat, and the cold assists one wonderfully in that operation. You consume a pound or so of beefsteaks at breakfast and then walk the deck for an appetite at dinner, when you take another pound or two of beef or a goose, or some such trifle. By four o'clock it is dark night, and as it is too cold to read, the only thing to be

---

[1] Captain Sulivan, who sailed with Darwin in the *Beagle*, and served with great distinction in command of the southern division of the fleet in the battle of Obligado (Plate River), had surveyed the Falkland Islands many years before his temporary settlement there. During the Crimean War he was surveying officer to the Baltic fleet, and afterwards naval adviser to the Board of Trade. He was afterwards Admiral and K.C.B.

done is to vanish under blankets as soon as possible and take twelve or fourteen hours' sleep.

Mrs. Stanley's Bougirigards,[1] which I have taken under my care during the cold weather, admire this sort of thing exceedingly and thrive under it, so I suppose I ought to.

The journey from New Zealand here has been upon the whole favourable ; no gales—quite the reverse—but light variable winds and calms. The latter part of our voyage has, however, been very cold, snow falling in abundance, and the ice forming great stalactites about our bows. We have seen no icebergs nor anything remarkable. From all I can learn it is most probable that we shall leave in about a week and shall go direct to England without stopping at any other port. I wish it may be so. I want to get home and look about me.

We have had news up to the end of March. There is nothing of any importance going on. By the Navy List for April I see that I shall be as nearly as possible in the middle of those of my own rank, i.e. I shall have about 150 above and as many below me. This is about what I ought to expect in the ordinary run of promotion in eight years, and I have served four and a half of that time. I don't expect much in the way of promotion, especially in these economic times ; but I do not fear that I shall be able to keep me in England for at least a year after our arrival, in order to publish my papers. The Admiralty have quite recently published a distinct declaration that they will consider scientific attainments as a claim to their notice, and I expect to be the first to remind them of their promise, and I will take care to have the reminder so backed that they must and shall take note of it. Even if they will not promote me at once, it would answer our purpose to have an appointment to some ship on the home station for a short time.

---

[1] The Australian love-bird ; a small parrakeet.

The last of the Falklands was seen on July 25 ; the line was crossed in thirty-six days; another month, and water running short, it was found necessary to put in at the Azores for a week. Leaving Fayal on October 5, the *Rattlesnake* reached Plymouth on the 23rd, but next day proceeded to Chatham, which, thanks to baffling winds, was not reached till November 9, when the ship was paid off.

CHAPTER V

1850-1851

In the Huxley Lecture for 1898 (*Times*, October 4) Professor Virchow takes occasion to speak of the effect of Huxley's service in the *Rattlesnake* upon his intellectual development :—

When Huxley himself left Charing Cross Hospital in 1846, he had enjoyed a rich measure of instruction in anatomy and physiology. Thus trained, he took the post of naval surgeon, and by the time that he returned, four years later, he had become a perfect zoologist and a keen-sighted ethnologist. How this was possible any one will readily understand who knows from his own experience how great the value of personal observation is for the development of independent and unprejudiced thought. For a young man who, besides collecting a rich treasure of positive knowledge, has practised dissection and the exercise of a critical judgment, a long sea voyage and a peaceful sojourn among entirely new surroundings afford an invaluable opportunity for original work and deep reflection. Freed from the formalism of the schools, thrown upon the use of his own intellect, compelled to test each single object as regards properties and history, he soon forgets the dogmas of the prevailing system and becomes, first a sceptic, and then an investigator. This change, which did not fail to affect Huxley, and through which arose that Huxley whom we commemorate to-day, is no unknown occurrence to one who is acquainted with the history, not only of knowledge, but also of scholars.

But he was not destined to find his subsequent path easy. Once in England, indeed, he did not lose any time. No sooner had the *Rattlesnake* touched at Plymouth than Commander Yule, who had succeeded Captain Stanley in the command of the ship, wrote to the head of the Naval Medical Department stating the circumstances under which Huxley's zoological investigations had been undertaken, and asking the sanction of the Admiralty for their publication. The hydrographer, in sending the formal permission, says :—

But I have to add that their Lordships will not allow any charge to be made upon the public funds towards the expense. You will, however, further assure Mr. Huxley that any assistance that can be supplied from this office shall be most cheerfully given to him, and that I heartily hope, from the capacity and taste for scientific investigation for which you give him credit, that he will produce a work alike creditable to himself, to his late Captain, by whom he was selected for it, and to Her Majesty's service.

Personally, the hydrographer took a great interest in science ; but as for the department, Huxley somewhat bitterly interpreted the official meaning of this well-sounding flourish to be made : "Publish if you can, and give us credit for granting every facility except the one means of publishing."

Happily there was another way of publishing, if the Admiralty would grant him time to arrange his papers and superintend their publication. The Royal Society had at their disposal an annual grant of money for the publication of scientific works. If the Government would not contribute directly to publish the researches made under their auspices, the favourable reception which his preliminary papers had met with led Huxley to hope that his greater work would be undertaken by the Royal Society. If the leading men of science attested the value of his work, the Admiralty might be induced to let him stay in England with the nominal appointment as assistant surgeon to H.M.S. *Fisguard* at Woolwich, for "particular service," but with leave of absence from the ship so that he could live and pursue his avocations in London. There was a precedent for this course in the case of Dr. Hooker, when he had to work out the scientific results of the voyage of the *Erebus* and *Terror*.

In this design he was fortified by his old Haslar friend, Dr. (afterwards Sir John) Watt Reid, who wrote : "They cannot, and, I am sure, will not wish to stand in your way at Whitehall." Meanwhile, the first person, naturally, he had thought of consulting was his old chief, Sir John Richardson, who had great weight at the Admiralty, and to him he wrote the following letter before leaving Plymouth :—

To Sir John Richardson

Oct. 31, 1850.

I regret very much that in consequence of our being ordered to be paid off at Chatham, instead of Portsmouth, as we always hoped and expected, I shall be unable to submit to your inspection the zoological notes and drawings which I have made during our cruise.   They are somewhat numerous (over 180 sheets of drawings), and I hope not altogether valueless, since they have been made with as great care and attention as I am master of—and with a microscope, such as has rarely, if ever, made a voyage round the world before.   A further reason for indulging in this hope consists in the fact that they relate for the most part to animals hitherto very little known, whether from their rarity or from their perishable nature, and that they bear upon many curious physiological points.

I may thus classify and enumerate the observations I have made—

1. Upon the organs of hearing and circulation in some of the transparent Crustacea, and upon the structure of certain of the lower forms of Crustacea.

2. Upon some very remarkable new forms of Annelids, and especially upon the much contested genus Sagitta, which I have evidence to show is neither a Mollusc nor an Epizoon, but an Annelid.

3. Upon the nervous system of certain Mollusca hitherto imperfectly described—upon what appears to me to be an urinary organ in many of them—and upon the structure of Firola and Atlanta, of which latter I have a pretty complete account.

4. Upon two perfectly new (ordinally new) species of Ascidians.

5. Upon Pyrosoma and Salpa.   The former has never been described (I think) since Savigny's time, and he had

only specimens preserved in spirits. I have a great deal
to add and alter. Then as to Salpa, whose mode of
generation has always been so great a bone of contention,
I have a long series of observations and drawings which
I have verified over and over again, and which, if correct,
must give rise to quite a new view of the matter. I may
mention as an interesting fact that in these animals so
low in the scale I have found a *placental circulation*,
rudimentary indeed, but nevertheless a perfect model on
a small scale of that which takes place in the mammalia.

6. I have the materials for a monograph upon the
Acalephæ and Hydrostatic Acalephæ. I have examined
very carefully more than forty genera of these animals—
many of them very rare, and some quite new. But I
paid comparatively little attention to the collection of
new species, caring rather to come to some clear and
definite idea as to the structure of those which had
indeed been long known, but very little understood.
Unfortunately for science, but fortunately for me, this
method appears to have been somewhat novel with
observers of these animals, and consequently everywhere
new and remarkable facts were to be had for the picking
up.

It is not to be supposed that one could occupy one's
self with the animals for so long without coming to some
conclusion as to their systematic place, however subsidiary to observation such considerations must always be
regarded, and it seems to me (although on such matters I
can of course only speak with the greatest hesitation) that
just as the more minute and careful observations made
upon the old "Vermes" of Linnæus necessitated the
breaking up of that class into several very distinct classes,
so more careful investigation requires the breaking up of
Cuvier's "Radiata" (which succeeded the "Vermes" as a
sort of zoological lumber-room) into several very distinct
and well-defined new classes, of which the Acalephæ,
Hydrostatic Acalephæ, actinoid and hydroid polypes, will

form one.   But I fear that I am trespassing beyond the limits of a letter.   I have only wished to state what I have done in order that you may judge concerning the propriety or impropriety of what I propose to do.   And I trust that you will not think that I am presuming too much upon your kindness if I take the liberty of thus asking your advice about my own affairs.   In truth, I feel in a manner responsible to you for the use of the appointment you procured for me ; and furthermore, Capt. Stanley's unfortunate decease has left the interests of the ship in general and my own in particular without a representative.

Can you inform me, then, what chance I should have either (1) of procuring a grant for the publication of my papers, or (2) should that not be feasible, to obtain a nominal appointment (say to the *Fisguard* at Woolwich, as in Dr. Hooker's case) for such time as might be requisite for the publication of my papers and drawings in some other way ?

I shall see Professors Owen and Forbes when I reach London, and I have a letter of introduction to Sir John Herschel (who has, I hear, a great penchant for the towing-net).   Supposing I could do so, would it be of any use to procure recommendations from them that my papers should be published ?

[(Half-erased) To Sir F. Beaufort also I have a letter.] Would it not be proper also to write to Sir W. Burnett acquainting him with my views, and requesting his acquiescence and assistance ?

Begging an answer at your earliest convenience, addressed either to the *Rattlesnake* or to my brother, I remain, your obedient servant,   T. H. HUXLEY.

41 North Bank.

He received a most friendly reply from "Old John."   He was willing to do all in his power to help, but could recommend Government aid better if he had seen the drawings. Meantime a certificate should be got from Forbes, the best man in this particular branch of science, backed, if possible, by Owen. He would speak to some officials himself, and give Huxley introductions to others, and if he could get up to town, would try to see the collections and add his name to the certificate.

Both Forbes and Owen were ready to help. The former wrote a most encouraging letter, singling out the characteristics which gave a peculiar value to these papers :—

I have had very great pleasure in examining your drawings of animals observed during the voyage of the *Rattlesnake*, and have also fully availed myself of the opportunity of going over the collections made during the course of the survey upon which you have been engaged. I can say without exaggeration that more important or more complete zoological researches have never been conducted during any voyage of discovery in the southern hemisphere. The course you have taken of directing your attention mainly to impreservable creatures, and to those orders of the animal kingdom respecting which we have least information, and the care and skill with which you have conducted elaborate dissections and microscopic examinations of the curious creatures you were so fortunate as to meet with, necessarily gives a peculiar and unique character to your researches, since thereby they fill up gaps in our knowledge of the animal kingdom. This is the more important, since such researches have been almost always neglected during voyages of discovery. The value of some of your notes was publicly acknowledged during your absence, when your memoir on the structure of the Medusæ, communicated to the Royal Society, was singled out for publication in the *Philosophical Transactions*. It would be a very great loss to science if the mass of new matter and fresh observation which you have accumulated were not to be worked out and fully published, as well as an injustice to the merits of the expedition in which you have served.

The latter offered to write to the Admiralty on his behalf, giving the weight of his name to the suggestion that the work to be done would take at least twelve months, and that therefore his appointment to the *Fisguard* should not be limited to any less period. "They might be disposed," wrote Huxley to him, "to cut anything I request down —on principle." Moreover, Owen, Forbes, Bell, and Sharpey, all members of the Committee of Recommendation of the Royal Society, had expressed themselves so favourably to his views, that in his application he was able to relieve the economic scruples of the Admiralty by telling them that he had a means of publishing his papers through the Royal Society.

The result of his application, thus backed, was that he obtained his appointment on November 29. It was for six months, subject to extension if he were able to report satisfactory progress with his work.

A long letter to his sister, now settled in Tennessee, gives a good idea of his aims and hopes at this time.

41 North Bank, Regent's Park,
November 21, 1850.

My dearest Lizzie—We have been at home now nearly three weeks, and I have been a free man again twelve days. Her Majesty's ships have been paid off on the 9th of this month. Properly speaking, indeed, we have been at home longer, for we touched at Plymouth and trod English ground and saw English green fields on the 23rd of October, but we were allowed to remain only twenty-four hours, and to my great disgust were ordered round to Chatham to be paid off. The ill-luck which had made our voyage homeward so long (we sailed from Sydney on the 2nd of May) pursued us in the Channel, and we did not reach Chatham until the 2nd of November; and what do you think was one of the first things I did when we reached Plymouth? Wrote to Eliza K. asking news of a certain naughty sister of mine, from whom I had never heard a word since we had been away—and if perchance there should be any letter, begging her to forward it immediately to Chatham. And so, when at length we got there, I found your kind long letter had been in England some six or seven months; but hearing of the likelihood of our return, they had very judiciously not sent it to me.

Your letter, my poor Lizzie, justifies many a heartache I have had when thinking over your lot, knowing, as I well do, what emigrant life is in climates less trying than that in which you live. I have seen a good deal of bush life in Australia, and it enables me fully to sympathise with and enter into every particular you tell me—from the baking and boiling and pigs squealing, down to that ferocious landshark Mrs. Gunther, of whose class Australia will furnish fine specimens. Had I been at home, too, I could have enlightened the good folks as to the means of carriage in the colonies, and could have told them that the two or twenty thousand miles over sea is the smallest

part of the difficulty and expense of getting anything to
people living inland ; as it is, I think I have done some
good in the matter ; their meaning was good but their
discretion small.   But the obtuseness of English in general
about anything out of the immediate circle of their own
experience is something wonderful.

I had heard here and there fractional accounts of your
doings from Eliza K. and my mother—not of the most
cheery description—and therefore I was right glad to get
your letter, which, though it tells of sorrow and misfortune enough and to spare, yet shows me that the brave
woman's heart you always had, my dearest Lizzie, is still
yours, and that you have always had the warm love of
those immediately around you, and now, as the doctor's
letter tells us, you have one more source of joy and
happiness, and this new joy must efface the bitterness—I
do not say the memory, knowing how impossible that
would be—of your great loss.[1]   God knows, my dear
sister, I could feel for you.   It was as if I could see again
a shadow of the great sorrow that fell upon us all years
ago.

Nothing can bind me more closely to your children
than I am already, but if the christening be not all over
you must let me be godfather ; and though I fear I am
too much of a heretic to promise to bring him up a good
son of the church—yet should ever the position which
you prophesy, and of which I have an "Ahnung"
(though I don't tell that to anybody but Nettie), be mine,
he shall (if you will trust him to me) be cared for as few
sons are.   As things stand, I am talking half nonsense,
but I mean it—and you know of old, for good and for
evil, my tenacity of purpose.

Now, as to my own affairs—I am not married.
Prudently, at any rate, but whether wisely or foolishly I
am not quite sure yet, Nettie and I resolved to have

---

[1] The death of her little daughter Jessie.

nothing to do with matrimony for the present. In truth, though our marriage was my great wish on many accounts, yet I feared to bring upon her the consequences that might have occurred had anything happened to me within the next few years. We had a sad parting enough, and as is usually the case with me, time, instead of alleviating, renders more disagreeable our separation. I have a woman's element in me. I hate the incessant struggle and toil to cut one another's throat among us men, and I long to be able to meet with some one in whom I can place implicit confidence, whose judgment I can respect, and yet who will not laugh at my most foolish weaknesses, and in whose love I can forget all care. All these conditions I have fulfilled in Nettie. With a strong natural intelligence, and knowledge enough to understand and sympathise with my aims, with firmness of a man, when necessary, she combines the gentleness of a very woman and the honest simplicity of a child, and then she loves me well, as well as I love her, and you know I love but few—in the real meaning of the word, perhaps, but two —she and you. And now she is away, and you are away. The worst of it is I have no ambition, except as means to an end, and that end is the possession of a sufficient income to marry upon. I assure you I would not give two straws for all the honours and titles in the world. A worker I must always be—it is my nature—but if I had £400 a year, I would never let my name appear to anything I did or shall ever do. It would be glorious to be a voice working in secret and free from all those personal motives that have actuated the best. But, unfortunately, one is not a "vox et præterea nihil," but with a considerable corporality attached which requires feeding, and so while my inner man is continually indulging in these anchorite reflections, the outer is sedulously elbowing and pushing as if he dreamed of nothing but gold medals and professors' caps.

I am getting on very well—better I fear than I

deserve. One of my papers was published in 1849 in the *Philosophical Transactions*, another in the *Zoological Transactions*, and some more may be published in the *Linnæan* if I like—but I think I shall not like. Then I have worked pretty hard, and brought home a considerable amount of drawings and notes about new or rare animals, all particularly nasty slimy things, and they will most likely be published as a separate work by the Royal Society.

Owen, Forbes, Bell, and Sharpey (the doctor will tell you of what weight these names are) are all members of the committee which disposes of the money, and are all strongly in favour of my "valuable researches" (cock-a-doodle-doo ! !) being published by the Society. From various circumstances I have taken a better position than I could have expected among these grandees, and I find them all immensely civil and ready to help me on, tooth and nail, particularly Prof. Forbes, who is a right good fellow, and has taken a great deal of trouble on my behalf. Owen volunteered to write to the "First Lord" on my behalf, and did so. Sharpey, when I saw him, reminded me, as he always does, of my great contest with Stocks[1] (do you remember throwing the shoe?) and promised me all the assistance in his power. Prof. Bell, who is secretary to the Royal, and has great influence, promised to help me in every way, and asked me to dine with him and meet a lot of nobs. I take all these things quite as a matter of course, but am all the while considerably astonished. The other day I dined at the Geological Club and met Lyell, Murchison, de la B[eche], Horner, and a lot more, and last evening I dined with a whole lot of literary and scientific people.

Owen was, in my estimation, great, from the fact of his smoking his cigar and singing his song like a brick.

I tell you all these things to show you clearly how I

---

[1] See p. 25.

stand. I am under no one's *patronage*, nor do I ever
mean to be. I have never asked, and I never will ask,
any man for his help from mere motives of friendship.
If any man thinks that I am capable of forwarding the
great cause in ever so small a way, let him just give me
a helping hand and I will thank him, but if not, he is
doing both himself and me harm in offering it, and if it
should be necessary for me to find public expression to
my thoughts on any matter, I have clearly made up my
mind to do so, without allowing myself to be influenced
by hope of gain or weight of authority.

There are many nice people in this world for whose
praise or blame I care not a whistle. I don't know and
I don't care whether I shall ever be what is called a great
man. I will leave my mark somewhere, and it shall be
clear and distinct [T. H. H., his mark.] and free from the
abominable blur of cant, humbug, and self-seeking which
surrounds everything in this present world—that is to
say, supposing that I am not already unconsciously tainted
myself, a result of which I have a morbid dread. I am
perhaps overrating myself. You must put me in mind
of my better self, as you did in your last letter, when
you write.

But I must come to the close of my epistle, as I have
one to enclose from my mother. My next shall be longer,
and I hope I shall then be able to tell you what I am
doing. At any rate I hope to be in England for twelve
months.

I am very much ashamed of myself for not having
written to you for so long—open confession is good for
the soul, they say, and I will honestly confess that I was
half puzzled, half piqued, and altogether sulky at your
not having answered my last letter containing my love
story, of which I wrote you an account before anybody.
You must not suppose my affection was a bit the less
because I was half angry. Nettie, who knows you well,
could tell you otherwise. Indeed, now that I know all,

I consider myself a great brute, and I will give you leave,
if you will but write soon, to scold me as much as you
like. All the family are well. My father is the only
one who is much altered, and that in mind and strength,
not in bodily health, which is very good. My mother
has lost her front teeth, but is otherwise just the same
amusing, nervous, distressingly active old lady she always
was.

Our cruisers visit New Orleans sometimes, and if ever
I am on the West India station, who knows, I may take
a run up to see you all. Kindest love to the children.
Tell Florry that I could not get her the bird with the
long tail, but that some day I will send her some pictures
of copper-coloured gentlemen with great big wigs and no
trousers, and tell her her old uncle loves her very much
and never forgets her nor anybody else.

God bless you, dearest Lizzie. Write soon.—Ever
your brother,   TOM.

Thus within a month of landing in England,
Huxley had secured his footing in the scientific
world. He was freed for the time from the more
irksome part of his profession; his service in the
navy had become a stepping-stone to the pursuits in
which his heart really was. He had long been half
in despair over the work which he had sent out like
the dove from the ark, if haply it might find him
some standing ground in the world; no news of it
had reached him till he was about to start on his
homeward voyage, but he returned to discover that
at a single stroke it had placed him in the front rank
of naturalists.

41 NORTH BANK, REGENT'S PARK,
Jan. 3, 1851.

My progress (he writes)[1] must necessarily be slow
and uncertain. I cannot see two steps forwards. Much
depends upon myself, much upon circumstances. Hitherto
all has gone as well as I could wish. I have gained each
object that I had set before myself—that is, I have my
shore appointment, I have found a means of publishing
what I have done creditably, and I have continued to
come into communication with some of the first men in
England in my department of science. But, as I have
found to be the case in all things that are gained, from
money to friendship, it is not so much getting as keeping.
It is by no means difficult if you are decently introduced,
have tolerably agreeable manners, and some smattering of
science, to take a position among these folks, but it is a
mighty different affair to keep it and turn it to account.
Not like the man who, at the Enchanted Castle, had the
courage to blow the horn but not to draw the sword, and
was consequently shot forth from the mouth of the cave
by which he entered with most ignominious haste,—one
must be ready to fight immediately after one's arrival has
been announced, or be blown into oblivion.

I *have* drawn the sword, but whether I am in truth to
beat the giants and deliver my princess from the enchanted
castle is yet to be seen.

For several months he lived with his brother
George and his wife at North Bank, St. John's Wood
(the house was pulled down in 1896 for the Great
Central Railway), but the surroundings were too
easy, and not conducive to hard work.

I must, I fear, emigrate to some "two pair back,"
which shall have the feel and manner of a workshop,

---

[1] When not otherwise specified, the extracts in this chapter are
from letters to his future wife.

where I can leave my books about and dissect a marine
nastiness if I think fit, sallying forth to meet the world
when necessary, and giving it no more time than necessary.
If it were not for a fear that P. would take it unkindly,
I should go at once.   I must summon up moral courage
somehow (how difficult when it is to pain those we love!)
and trust to her good sense for the rest.

And later : —

. . . I have been very busy looking about for the last
two days, and have been in fifty houses if I have been in
one.   I want some place with a decent address, cheap, and
beyond all things, clean.   The dirty holes that some of
these lodgings are ! such tawdry finery and such servants,
with their faces and hands not merely dirty, but absolutely
macadamised.   And they all make this confounded great
Exhibition a plea for about doubling the rent.

So in April 1851 he removed to lodgings hard by,
at 1 Hanover Place, Clarence Gate, Regent's Park
(" which sounds grand, but means nothing more than
a sitting-room and bedroom in a small house "), then
to St. Anne's Gardens, and after that to Upper York
Place, while making a second home with his brother.
His other great friends already in London were the
Fannings, who had left Australia a few months before
his own return.   In the scientific world he soon made
acquaintance with most of the leading men, and began
a close friendship with Edward Forbes, with George
Busk (then surgeon to H.M.S. *Dreadnought* at
Greenwich, afterwards President of the College of
Surgeons) and his accomplished wife, and later in the
year with both Hooker and Tyndall.   The Busks,

indeed, showed him the greatest kindness throughout this period of struggle, and the sympathy and intellectual stimulus he received from their society were of the utmost help. They were always ready to welcome him at Greenwich, and he not only often ran down there for a week-end, but would spend part of his vacations with them at Lowestoft or Tenby, where naturalists could find plenty of occupation.

But from a worldly point of view, it was too soon clear that science was sadly unprofitable. There seemed no speedy prospect of making enough to marry on. As early as March 1851 he writes :—

The difficulties of obtaining a decent position in England in anything like a reasonable time seem to me greater than ever they were. To attempt to live by any scientific pursuit is a farce. Nothing but what is absolutely practical will go down in England. A man of science may earn great distinction, but not bread. He will get invitations to all sorts of dinners and conversaziones, but not enough income to pay his cab fare. A man of science in these times is like an Esau who sells his birthright for a mess of pottage. Again, if one turns to practice, it is still the old story—wait : and only after years of working like a galley-slave and intriguing like a courtier is there any chance of getting a decent livelihood. I am not at all sure if . . . it would not be the most prudent thing to stick by the Service : there at any rate is certainty in health and in sickness.

Nevertheless he was mightily encouraged in the work of bringing out his *Rattlesnake* papers by a notable success in a quarter where he scarcely dared to hope for it. The Royal Society had for some time set itself to become a body of working men of science; to exclude for the future all mere dilettanti, and to admit a limited number of men whose work was such as to deserve recognition. Thanks to the initiative of Forbes, he now found this recognition accorded to him on the strength of his "Medusa" paper. He writes in February :—

> The F.R.S. that you tell me you dream of being appended to my name is nearer than one might think, to my no small surprise. . . . I had no idea that it was at all within my reach, until I found out the other day, talking with Mr. Bell, that my having a paper in the *Transactions* was one of the best of qualifications.
>
> My friend Forbes, to whom I am so much indebted, has taken the matter in hand for me, and I am told I am sure of getting it this year or the next. I do not at all expect it this year, as there are a great many candidates, far better men than I. . . . I shall think myself lucky if I get it next year. Don't say anything about the matter till I tell you. . . . As the old proverb says, there is many a slip 'twixt the cup and the lip.

There were thirty-eight candidates; of these the Council would select fifteen, and submit their names for election at a general meeting of the Society. He was not yet twenty-six years of age, and certainly the youngest and least known of the competitors. Others probably had been up before — possibly many times before; nevertheless, on this, his first candidature, he was placed among the selected. The formal election did not take place till June 5, but on a chance visit to Forbes he heard the great

news. The F.R.S. was a formal attestation of the
value of the work he had already done; it was a
token of success in the present, an augury of greater
success in the future. No wonder the news was
exciting.

To-day (he writes on April 14) I saw Forbes at the
Museum of Practical Geology, where I often drop in on
him. "Well," he said, "I am glad to be able to tell you
you are all right for the Royal Society; the selection was
made on Friday night, and I hear that you are one of
the selected. I have not seen the list, but my authority
is so good that you may make yourself easy about it."
I confess to having felt a little proud, though I believe
I spoke and looked as cool as a cucumber. There were
thirty-eight candidates, out of whom only fifteen could
be selected, and I fear that they have left behind much
better men than I. I shall not feel certain about the
matter until I receive some official announcement. I
almost wish that until then I had heard nothing about
it. Notwithstanding all my cucumbery appearance, I
will confess to you that I could not sit down and read
to-day after the news. I wandered hither and thither
restlessly half over London. . . . Whether I have it or
not, I can say one thing, that I have left my case to stand
on its own strength; I have not asked for a single vote,
and there are not on my certificate half the names that
there might be. If it be mine, it is by no intrigue.

Again, on May 4 :—

I am twenty-six to-day . . . and it reminds me that
I have left you now a whole year. It is perfectly
frightful to think how the time is slipping by, and yet
seems to bring us no nearer.
What have I done with my twenty-sixth year? Six

months were spent at sea, and therefore may be considered
as so much lost; and six months I have had in England.
That, I may say, has not been thrown away altogether
without fruit. I have read a good deal and I have
written a good deal. I have made some valuable friends,
and have found my work more highly estimated than
I had ventured to hope. I must tell you something,
because it will please you, even if you think me vain
for doing so.

I was talking to Professor Owen yesterday, and said
that I imagined I had to thank him in great measure for
the honour of the F.R.S. "No," he said, "you have
nothing to thank but the goodness of your own work."
For about ten minutes I felt rather proud of that speech,
and shall keep it by me whenever I feel inclined to think
myself a fool, and that I have a most mistaken notion of
my own capacities. The only use of honours is as an
antidote to such fits of the "blue devils." Of one thing,
however, which is by no means so agreeable, my opportunities for seeing the scientific world in England force
upon me every day a stronger and stronger conviction.
It is that there is no chance of living by science. I have
been loth to believe it, but it is so. There are not more
than four or five offices in London which a Zoologist or
Comparative Anatomist can hold and live by. Owen,
who has a European reputation, second only to that of
Cuvier, gets as Hunterian Professor £300 a year! which
is less than the salary of many a bank clerk. My friend
Forbes, who is a highly distinguished and a very able
man, gets the same from his office of Palæontologist to
the Geological Survey of Great Britain. Now, these are
first-rate men—men who have been at work for years
laboriously toiling upward—men whose abilities, had they
turned them into the many channels of money-making,
must have made large fortunes. But the beauty of
Nature and the pursuit of Truth allured them into a
nobler life—and this is the result. . . . In literature a

man may write for magazines and reviews, and so support himself ; but not so in science.   I could get anything I write into any of the journals or any of the Transactions, but I know no means of thereby earning five shillings. A man who chooses a life of science chooses not a life of poverty, but, so far as I can see, a life of *nothing*, and the art of living upon nothing at all has yet to be discovered. You will naturally think, then, " Why persevere in so hopeless a course ?"   At present I cannot help myself. For my own credit, for the sake of gratifying those who have hitherto helped me on—nay, for the sake of truth and science itself, I must work out fairly and fully complete what I have begun.   And when that is done, I will courageously and cheerfully turn my back upon all my old aspirations.   The world is wide, and there is everywhere room for honesty of purpose and earnest endeavour.   Had I failed in attaining my wishes from an overweening self-confidence,—had I found that the obstacles after all lay within myself—I should have bitterly despised myself, and, worst of all, I should have felt that you had just ground of complaint.

So far as the acknowledgment of the value of what I have done is concerned, I have succeeded beyond my expectations, and if I have failed on the other side of the question, I cannot blame myself.   It is the world's fault and not mine.

A few months more, and he was able to report another and still more unexpected testimony to the value of his work—another encouragement to persevere in the difficult pursuit of a scientific life.   He found himself treated as an equal by men of established reputation ; and the first-fruits of his work ranked on a level with the maturer efforts of veterans in science.   He was within an ace of receiving the

Royal Medal, which was awarded him the following year. Of this he writes :—

November 7, 1851.—I have at last tasted what it is to mingle with my fellows—to take my place in that society for which nature has fitted me, and whether the draught has been a poison which has heated my veins, or true nectar from the gods, life-giving, I know not, but I can no longer rest where I once could have rested. If I could find within myself that mere peisonal ambition, the desire of fame, present or posthumous, had anything to do with this restlessness, I would root it out. But in those moments of self-questioning, when one does not lie even to one's self, I feel that I can say it is not so—that the real pleasure, the true sphere, lies in the feeling of self-development—in the sense of power and of growing *oneness* with the great spirit of abstract truth.

Do you understand this? I know you do; our old oneness of feeling will not desert us here. . . .

To-day a most unexpected occurrence came to my knowledge. I must tell you that the Queen places at the disposal of the Royal Society once a year a valuable gold medal to be given to the author of the best paper upon either a physical, chemical, or anatomical or physiological subject. One of these branches of science is chosen by the Royal Society for each year, and therefore for any given subject—say anatomy and physiology—it becomes a triennial prize, and is given to the best memoir in the *Transactions* for three years.

It happens that the Royal Medal, as it is called, is this year given in Anatomy and Physiology. I had no idea that I had the least chance of getting it, and made no effort to do so. But I heard this morning from a member of the Council that the award was made yesterday, and that I was within an ace of getting it. Newport,[1] a man

---

[1] George Newport, 1803-1854. His most important work was that on "Impregnation," for which he received this medal.

of high standing in the scientific world, and myself were the two between whom the choice rested, and eventually it was given to him, on account of his having a greater bulk of matter in his papers, so evenly did the balance swing. Had I only had the least idea that I should be selected they should have had enough and to spare from me. However, I do not grudge Newport his medal; he is a good and a worthy competitor, old enough to be my father, and has long had a high reputation. Except for its practical value as a means of getting a position I care little enough for the medal. What I do care for is the justification which the being marked in this position gives to the course I have taken. Obstinate and self-willed as I am . . . there are times when grave doubts overshadow my mind, and then such testimony as this restores my self-confidence.

To let you know the full force of what I have been saying, I must tell you that this "Royal Medal" is what such men as Owen and Faraday are glad to get, and is indeed one of the highest honours in England.

To-day I had the great pleasure of meeting my old friend Sir John Richardson (to whom I was mainly indebted for my appointment in the *Rattlesnake*). Since I left England he has married a third wife, and has taken a hand in joining in search of Franklin (which was more dreadful ?), like an old hero as he is; but not a feather of him is altered, and he is as grey, as really kind, and as seemingly abrupt and grim, as ever he was. Such a fine old polar bear !

CHAPTER VI

1851–1854

THE course pursued by the Government in the matter of Huxley's papers is curious and instructive. The Admiralty minute of 1849 had promised either money assistance for publishing or speedy promotion as an encouragement to scientific research in the Navy, especially by the medical officers. On leave to publish the scientific results of the expedition being asked for, the Department forestalled any request for monetary aid by an intimation that none would be given. Strong representations, however, from the leading scientific authorities induced them to grant the appointment to the *Fisguard* for six months.

The sequel shows how the departmental representatives of science did their best for science in Huxley's case, so far as in their power lay :—

June 6, 1851.—The other day I received an intimation that my presence was required at Somerset House. I rather expected the mandate, as my six months' leave was up. Sir William was very civil, and told me that the Commander of the *Fisguard* had applied to the

Admiralty to know what was to be done with me, as my leave had expired. "Now," said he, "go to Forrest" (his secretary), "write a letter to me, stating what you want, and I will get it done for you." So away I went and applied for an indefinite amount of leave, on condition of reporting the progress of my work every six months, and as I suppose I shall get it, I feel quite easy on that head.

In May 1851 he applied to the Royal Society for help from the Government Grant towards publishing the bulk of his work as a whole, for much of its value would be lost if scattered fragmentarily among the Transactions of various learned societies. Personally, the members of the committee were very willing to make the grant, but on further consideration it appeared that the money was to be applied for promoting research, not for assisting publication; and moreover, it was desirable not to establish a precedent for saddling the funds at the disposal of the Society with all the publications which it was the clear duty of the Government to undertake. On this ground the application was refused, but at the same time it was resolved that the Government be formally asked to give the necessary subvention towards bringing out these valuable papers.

A similar resolution was passed at the Ipswich meeting of the British Association in July 1851, and at a meeting of its Council in March 1852 the President declared himself ready to carry it into effect by asking the Treasury for the needful £300. But at the July meeting he could only report a *non possumus* answer for the current year (1852) from the Government, and a resolution was passed recommending that application on the subject be renewed by the British Association in the following year.

Meanwhile, weary of official delay, Huxley had conceived the idea of writing direct to the Duke of Northumberland, then First Lord of the Admiralty, whom he knew to take an interest in scientific research. At the same time he stirred Lord Rosse, the President of the Royal Society, to repeat his application to the Treasury. Although the Admiralty in April 1852 again refused money help, and bade him apply to the Royal Society for a portion of the Government Grant (which the latter had already refused him), the Hydrographer was directed to make inquiries as to the propriety of granting him an extension of leave. To his question asking the exact amount of time still required for finishing the work of publication, Huxley returned what he described as a "savage reply," that his experience of engravers led him to think that the plates could be published in eight or nine months from the receipt of a grant; that he had reason to believe this grant might soon be promised, but that the long delay was solely due to the remissness of those whose duty it was to represent his claims to the Government; and finally, that he must ask for a year's extension of leave.

For these expressions his conscience smote him when, on June 12, at a soiree of the Royal Society,

Lord Rosse took him aside and informed him that he had seen Sir C. Trevelyan, the Under Secretary to the Treasury, who said there would be no difficulty in the matter if it were properly laid before the Prime Minister, Lord Derby. To Lord Derby therefore he went, and was told that Mr. Huxley should go to the Treasury and arrange matters in person with Trevelyan. At the same time the indignant tone of his letter to the Hydrographer seemed to have done good; he was invited to explain matters in person, and was granted the leave he asked for.

Everything now seemed to point to a speedy solution of his difficulties. The promise of a grant, of course, did nothing immediate, but assured him a good position, and settled all the scruples of the Admiralty with regard to time. "You have no notion," he writes, "of the trouble the grant has cost me. It died a natural death till I wrote to the Duke in March, and brought it to life again. The more opposition there is, the more determined I am to carry it through." But he was doomed to a worse disappointment than before. Trevelyan received him very civilly, but had heard nothing on the matter from Lord Derby, and accordingly sent him in charge of his private secretary to see Lord Derby's secretary. The latter had seen no papers relating to any such matter, and supposed Lord Derby had not brought them from St. James's Square, "but promised to write to me as soon as anything was learnt. I look upon it as adjourned *sine die.*" Parliament breaking up immediately after gave the officials a good excuse for doing nothing more.

When his year's leave expired in June 1853, he wrote the following letter to Sir William Burnett :—

As the period of my leave of absence from H.M.S. *Fisguard* is about to expire, I have the honour to report that the duty on which I have been engaged has been carried out, as far as my means permit, by the publication of a "Memoir upon the Homologies of the Cephalous Mollusca," with four plates, which appeared in the *Philosophical Transactions* for 1852 (published 1853), being the fourth memoir resulting from the observations made during the voyage of H.M.S. *Rattlesnake* which has appeared in these *Transactions*.

I have the pleasure of being able to add that the President and Council of the Royal Society have considered these memoirs worthy of being rewarded by the Royal Medal in Physiology for 1852, which they did me the honour to confer in the November of that year.

I regret that no definite answer of any kind having as yet been given to the strong representations which were made by the Presidents both of the Royal Society and of the British Association in 1852 to H.M. Government— representations which have recently been earnestly repeated—in order to obtain a grant for the purpose of publishing the remainder of these researches in a separate form, I have been unable to proceed any further, and I beg to request a renewal of my leave of absence from H.M.S. *Fisguard,* so that if H.M. Government think fit to give the grant applied for, it may be in my power to make use of it; or that, should it be denied, I may be enabled to find some other means of preventing the total loss of the labour of some years.

Hereupon he was allowed six months longer, but

with the intimation that no further leave would be granted. A final application from the scientific authorities resulted in fresh inquiries as to the length of time still required, and the deadlock between the two departments of State being unchanged, he replied to the same effect as before, but to no purpose. His formal application for leave in January 1854 was met by orders to join the *Illustrious* at Portsmouth. He appealed to the Admiralty that this appointment might be cancelled, giving a brief summary of the facts, and pointing out that it was the inaction of the Treasury which had absolutely prevented him from completing his work.

I would therefore respectfully submit that, under these circumstances, my request to be permitted to remain on half-pay until the completion of the publication of the results of some years' toil is not wholly unreasonable. It is the only reward for which I would ask their Lordships, and indeed, considering the distinct pledge given in the minute to which I have referred, to grant it would seem as nearly to concern their Lordships' honour as my advantage.

The counter to this bold stroke was crushing, if not convincing. He was ordered to join his ship immediately under pain of being struck off the Navy list. He was of course prepared for this ultimatum, and whether he could manage to pursue science in England or might be compelled to set up as a doctor in Sydney, he considered that he would be better off than as an assistant-surgeon in the Navy. Accordingly he stood firm, and the threat was carried into

effect in March 1854. An unexpected consequence
followed. As long as he was in the Navy, with
direct claims upon a Government department for
assistance in publishing his work, the Royal Society
had not felt justified in allotting him any part of the
Government Grant. But now that he had left the
service, this objection was removed, and in June
1854 the sum of £300 was assigned for this purpose,
while the remainder of the expense was borne by the
Ray Society, which undertook the publication under
the title of *Oceanic Hydrozoa*. Thus he was able to
record with some satisfaction how he at last has got
the grant, though indirectly, from the Government, and considers it something of a triumph
for the principle of the family motto, *tenax propositi.*

While these fruitless negotiations with the Admiralty were in progress, he had done a good deal, both
in publishing what he could of his *Rattlesnake* work,
and in trying to secure some scientific appointment
which would enable him to carry out his two chief
objects: the one his marriage, the other the un-
hampered pursuit of science. In addition to the
papers sent home from the cruise—one on the
Medusæ, published in the *Philosophical Transactions
of the Royal Society* for 1849, and one on the Animal
of Trigonia, published in the *Proceedings of the
Zoological Society* for the same year—he had reported
to the Admiralty in June 1851 the publication of
seven memoirs :—

1. On the Auditory Organs of the Crustacea. Published in the *Annals of Natural History*.

2. On the Anatomy of the genus Tethea. Published in the *Annals of Natural History*.

3. Report upon the Development of the Echinoderms. To appear in the *Annals* for July.

4. On the Anatomy and Physiology of the Salpæ, with four plates. Read at the Royal Society, and to be published in the next part of the *Philosophical Transactions*.

5. On two Genera of Ascidians, Doliolum and Appendicularia, with one plate. Read at the Royal Society, and to be published in the next part of the *Philosophical Transactions*.

6 On some peculiarities in the Circulation of the Mollusca. Sent to M. Milne-Edwards, at his request, to be published in the *Annales des Sciences*.

7. On the Generative Organs of the Physophoridæ and Diphydæ. Sent to Prof. Muller of Berlin for publication in his *Archiv*.

By the end of the year he had four more to report:—1. On the Hydrostatic Acalephæ; 2. On the genus Sagitta, both published in the *Report of the British Association* for 1851; 3. On Lacinularia Socialis, a contribution to the anatomy and physiology of the Rotifera, in the *Transactions of the Microscopical Society*. 4. On Thalassicolla, a new zoophyte, in the *Annals of Natural History*. Next year he read before the British Association a paper entitled "Researches into the Structure of the Ascidians," and a very important one on the Morphology of the Cephalous Mollusca, afterwards published in the *Philosophical Transactions*. In addition he had prepared a great part of his longer work for publication; out of twenty-four or twenty-five plates, nineteen were ready for the engraver when he wrote his appeal to the Duke of Northumberland. In this same year, 1852, he was also awarded the Royal Medal in Physiology for the value of his contributions to the *Philosophical Transactions*.

In 1853, besides seeing some of these papers through the press, he published one on the existence of Cellulose in the Tunic of Ascidians, read before the Microscopical Society, and two papers on the Structure of the Teeth; the latter, of course, like a paper of the previous year on Echinococcus, being distinct from the *Rattlesnake* work. The greater work on Oceanic Hydrozoa, over which the battle of the grant in aid had been waged so long, did not see the light until 1858, when his interest had been diverted from these subjects, and to return to them was more a burden than a pleasure.

In the second place, the years 1851-53, so full of profitless successes in pure science, and delusive hopes held out by the Government, were marked by an equally unsuccessful series of attempts to obtain a professorship. If a chair of Natural History had been established, as he hoped, in the projected university at Sydney, he would gladly have stood for it. Sydney was a second home to him; he would have been backed by the great influence of Macleay; and in his eyes a naturalist could not desire a finer field for his labours than the waters of Port Jackson.

But this was not to be, and the first chair he tried for was the newly-instituted chair of Zoology at the University of Toronto. The vacancy was advertised in the summer of 1851; the pay of full £300 a year was enough to marry on; his friends reassured him as to his capacity to fill the post, which, moreover, did not debar him from the hope of returning some day to fill a similar post in England.

1 EDWARD STREET, ST. JOHN'S WOOD TERRACE,
July 29 [1851].

MY DEAR HENFREY—I have been detained in town, or I hope we should long since have had our projected excursion.

What do you think of my looking out for a Professorship of Natural History at Toronto? Pay £350, with chances of extra fees. I think that out there one might live comfortably upon that sum—possibly even do the domestic and cultivate the Loves and Graces as well as the Muses.

Seriously, however, I should like to know what you think of it. The chance of getting anything over here without devoting one's self wholly to Mammon, seems to me very small. At least it involves years of waiting.

Toronto is not very much out of the way, and the pay is decent and would enable me to devote myself wholly to my favourite pursuits. Were it in England, I could wish nothing better; and, as it is, I think it would answer my purpose very well for some years at any rate.

If they go fairly to work, I think I shall have a very good chance of being elected; but I am told that these matters are often determined by petty intrigues.

Francis [1] and I looked for you everywhere at the

---

[1] Dr. William Francis, one of the editors of the *Philosophical Magazine*, and a member of the publishing firm of Taylor and Francis.

Botanic Gardens, and finding you were too wise to come, came here, grieving your absence, and had an æsthetic " Bier."

He obtained a remarkably strong set of testimonials from all the leading anatomists and physiologists in the kingdom, as well as one from Milne-Edwards in Paris.

I have put together (he writes) twelve or fourteen testimonials from the first men.   I will have no other.

His newly-obtained F.R.S. was a recommendation in itself.   So that he writes :—

There are, I learn, several other candidates, but no one I fear at all, if they only have fair play.   There is no one of the others who can command anything like the scientific influence which is being exercised for me, whatever private influence they may have.

What makes all the big-wigs so marvellously zealous on my behalf I know not.   I have sought none of them and flattered none of them, that I can say with a good conscience, and I think you know me well enough to believe it.   I feel very grateful to them ; and if it ever happens that I am able to help a young man on (when I am a big-wig myself !) I shall remember it.

And again, September 23, 1851 :—

When I have once sent away my testimonials and done all that is to be done, I shall banish the subject from my mind and make myself quite easy as to results For the present I confess to being somewhat anxious.

Nevertheless, after many postponements, a near relative of an influential Canadian politician was at length appointed late in 1853.   By an amusing coincidence, Huxley's newly-made friend, Tyndall, was likewise a candidate for a chair at Toronto, and likewise rejected. Two letters, concerning Tyndall's election to the Royal Society, contain references both to Toronto and to Sydney.

4 UPPER YORK PLACE, ST. JOHN'S WOOD,
Dec. 4 [1851].

MY DEAR SIR—I was greatly rejoiced to find I could be of service to you in any way, and I only regret, for your sake, that my name is not a more weighty one. Your election, I should think, can be a matter of no doubt.

As to Toronto, I confess I am not very anxious about it. Sydney would have been far more to my taste, and I confess I envy you what, as I hear, is the very good chance you have of going there.

It used to be our headquarters in the *Rattlesnake* and my home for three months in the year. Should you go, I should be very happy, if you like, to give you letters to some of my friends.

Greatly as I wish we had been destined to do our work together, I cannot but offer the most hearty wishes for your success in Sydney.—Ever yours very faithfully

THOMAS H. HUXLEY.

John Tyndall, Esq.

41 NORTH BANK, REGENT'S PARK,
May 7, 1852.

MY DEAR TYNDALL—Allow me to be one of the first to have the pleasure of congratulating you on your new honours. I had the satisfaction last night to hear your name read out as one of the selected of the Council of the Royal Society for election to the Fellowship this year, and you are therefore as good as elected.

I always made sure of your success, but I am not the

less pleased that it is now a *fait accompli.*—I am, my dear Tyndall, faithfully yours,   T. H. HUXLEY.

P.S.—I have heard nothing of Toronto, and I begin to think that the whole affair, University and all, is a myth.

His hopes of the Colonies failing, he tried each of the divisions of the United Kingdom in turn, with uniform ill-success; in 1852-53 at Aberdeen and at Cork; in 1853 at King's College, London. He had great hopes of Aberdeen at first; the appointment lay with the Home Secretary, a personal friend of Sir J. Clark, who was interested in Huxley though not personally acquainted with him. But no sooner had he written to urge the latter's claims than a change of ministry took place, and other influences commanded the field. It was cold comfort that Clark told him only to wait—something must turn up. There was still a great probability of the Toronto chair falling to a Cork professor; so with this in view, he gave up a trip to Chamounix with his brother, and attended the meeting of the British Association at Belfast in August 1852, in order to make himself known to the Irish men of science, for, as his friends told him, personal influence went for so much, and while most men's reputations were better than themselves, he might flatter himself that he was better than his reputation. But this, too, came to nothing, and the King's College appointment also went to the candidate who was backed by the most powerful influence.

A fatality seemed to dog his efforts; nevertheless he writes at the end of 1851 :—

Among my scientific friends the monition I get on all sides is that of Dante's great ancestor to him—

Se tu segui la tua stella.

If this were from personal friends only, I should disregard it; but it comes from men to whose approbation it would be foolish affectation to deny the highest value. I find myself treated on a footing of equality ("my proud self," as you may suppose, would not put up with any other) by men whose names and works have been long before the world. My opinions are treated with a respect altogether unaccountable to me, and what I have done is quoted as having full authority. Without canvassing a soul or making use of any influence, I have been elected into the Royal Society at a time when that election is more difficult than it has ever been in the history of the Society. Without my knowledge, I was within an ace of getting the Royal Society medal this year, and if I go on I shall very probably get it next time.

In 1852 he was not only to receive this coveted honour,[1] but also to be elected upon the Royal Society Council. In January 1852, when standing for Toronto, he describes how Col. Sabine, then Secretary of the Royal Society, dissuaded him from the project, saying that a brilliant prospect lay before him if he would only wait.

"Make up your mind to get something fairly within your reach, and you will have us all with you." Prof. Owen again offers to do anything in his power for me;

---

[1] See pp. 149 *sqq.*

Prof. Forbes will move heaven and earth for me if he can ; Gray, Bell, and all the leading men are, I know, similarly inclined. Fate says wait, and you shall reach the goal which from a child you have set before your-self. On the other hand, a small voice like conscience speaks of one who is wasting youth and life away for your sake.

Other friends, who, while recognising his general capacities, were not scientific, and had no direct appreciation of his superlative powers in science, thought he was following a course which would never allow him to marry, and urged him to give up this un-equal battle with fate, and emigrate to Australia. Of this he writes on August 5, 1852, to Miss Heathorn :—

I must make up my mind to it if nothing turns up. However, I look upon such a life as would await me in Australia with great misgiving. A life spent in a routine employment, with no excitement and no occupation for the higher powers of the intellect, with its great aspira-tions stifled and all the great problems of existence set hopelessly in the background, offers to me a prospect that would be utterly intolerable but for your love. . . . Sometimes I am half mad with the notion of burying all my powers in a mere struggle for a livelihood. Some-times I am equally wild at thinking of the long weary while that has passed since we met. There are times when I cannot bear to think of leaving my present pursuits, when I feel I should be guilty of a piece of cowardly desertion from my duty in doing it, and there come intervals when I would give truth and science and all hopes to be folded in your arms. . . . I know which course is right, but I never know which I may follow ; help me . . . for there is only one course in which there is either hope or peace for me.

These repeated disappointments deepened the fits of depression which constantly assailed him. He was torn by two opposing thoughts. Was it just, was it right, to demand so great a sacrifice from the woman who had entrusted her future to the uncertain chances of his fortunes? Could he ask her to go on offering up the best years of her life to aspirations of his which were possibly chimerical, or perhaps merely selfishness in disguise, which ought to yield to more imperative duties? Why not clip the wings of Pegasus, and descend to the sober, everyday jog-trot after plain bread and cheese like other plain people? Time after time he almost made up his mind to throw science to the winds; to emigrate and establish a practice in Sydney; to try even squatting or storekeeping. And yet he knew only too well that with his temperament no life would bring him the remotest approach to lasting happiness and satisfaction except one that gave scope to his intellectual passion. To yield to the immediate pressure of circumstances was perhaps ignoble, was even more probably a surer road to the loss of happiness for himself and for his wife than the repeated and painful sacrifices of the present. With all this, however, and the more when assured of her entire confidence in his judgment, he could not but feel a sense of remorse that she willingly accepted the sacrifice, and feared that she might have done so rather to gratify his wishes than because reason approved it as the right course to follow.

Here is another typical extract from his correspondence. Hearing that Toronto is likely to go to a relative of a Canadian minister, he writes, January 2, 1852 :—

I think of all my dreams and aspirations, and of the path which I know lies before me if I can only bide my time, and it seems a sin and a shameful thing to allow my resolve to be turned ; and then comes the mocking suspicion, is this fine abstract duty of yours anything but a subtlety of your own selfishness ? Have you not other more imperative duties ?

You may fancy whether my life is a very happy one thus spent without even the satisfaction of the sense of right-doing. I must come to some resolution about it, and that shortly. I was talking seriously with Fanning the other night about the possibility of finding some employment of a profitable kind in Australia, store-keeping, squatting, or the like. As I told him, any change in my mode of life must be *total*. If I am to change at all, the change must be total and complete. I will not attempt my own profession. I should only be led astray to think and to work as of old, and sigh continually for my old dear and intoxicating pursuits. I wish I understood Brewing, and I would make a proposition to come and help your father. You may smile, but I am as serious as ever I was in my life.

The distance between them made it doubly difficult to keep in touch with one another, when the post took from four and a half to five or even six months to reach England from Australia. The answer to a letter would come when the matter in question was long done with. The assurance that he was doing right at one moment seemed inadequate when circumstances had altered and hope sunk lower. It was all too easy to suspect that she did not understand his aims, his thirst for action, nor the fact that he was no longer free to do as he liked, whether to stay in the navy, to go into practice, or follow his own pursuits and pleasure. Yet it made him despair to be so hedged in by circumstances. With all his efforts, he seemed as though he had done nothing but earn the reputation of being a very promising young man. How much easier to continue the struggle if he could but have seen her face to face, and read her thoughts as to whether he were right or wrong in the course he was pursuing. He appeals to her faith that he is choosing the nobler path in pursuing knowledge, than in turning aside to the temptation of throwing it up for the sake of their speedier union. Still she was right in claiming a share in his work; but for her his life would have been wasted.

The clouds gathered very thickly about him when in April 1852 his mother died, while his father was hopelessly ill. "Belief and happiness," he writes, "seem to be beyond the reach of thinking men in these days, but courage and silence are left." Again the clouds lifted, for in October he received Miss Heathorn's "noble and self-sacrificing letter, which has given me more comfort than anything for a long while," the keynote of which was that a man should pursue those things for which he is most fitted, let them be what they will. He now felt free to tell

the vicissitudes of thought and will he had passed
through this twelvemonth, and how the idea of giving
up all had affected him.   " The spectre of a wasted
life has passed before me—a vision of that servant
who hid his talent in a napkin and buried it."

Early in 1853 he writes how much he was cheered
by his sister's advice and encouragement to persist in
the struggle; but the darkest moment was still to
come. His hopes from his candidatures crumbled away
one after the other; his leave from the Admiralty
was coming to an end, and there was small hope of
renewing it; the grant from Government remained
as unattainable as ever; the long struggle had taught
him the full extent of his powers only, it seemed, to
end by denying him all opportunity for their use.

And so the card house I have been so laboriously
building up these two years with all manner of hard
struggling will be tumbled down again, and my small
light will be ignominiously snuffed out like that of better
men. . . . I can submit if the fates are too strong.   The
world is no better than an arena of gladiators, and I, a
stray savage, have been turned into it to fight my way
with my rude club among the steel-clad fighters.   Well,
I have won my way into the front rank, and ought to be
thankful and deem it only the natural order of things if
I can get no further.

And again in a letter of July 6, 1853 :—

I know that these three years have inconceivably
altered me—that from being an idle man, only too happy
to flow into the humours of the moment, I have become
almost unable to exist without active intellectual excitement.   I know that in this I find peace and rest such

as I can attain in no other way. From being a mere untried fledgling, doubtful whether the wish to fly proceeded from mere presumption or from budding wings, I have now some confidence in well-tried pinions, which have given me rank among the strongest and foremost. I have always felt how difficult it was for you to realise all this—how strange it must be to you that though your image remained as bright as ever, new interests and purposes had ranged themselves around it, and though they could claim no pre-eminence, yet demanded their share of my thoughts. I make no apology for this—it is man's nature and the necessary influence of circumstances which will so have it; and depend, however painful our present separation may be, the spectacle of a man who had given up the cherished purpose of his life, the Esau who had sold his birthright for a mess of pottage and with it his self-respect, would before long years were over our heads be infinitely more painful. Depend upon it, the trust which you placed in my hands when I left you—to choose for both of us—has not been abused. Hemmed in by all sorts of difficulties, my choice was a narrow one, and I was guided more by circumstances than my own free will. Nevertheless the path has shown itself to be a fair one, neither more difficult nor less so than most paths in life in which a man of energy may hope to do much if he believes in himself, and is at peace within.

My course in life is taken. I will *not* leave London —I *will* make myself a name and a position as well as an income by some kind of pursuit connected with science, which is the thing for which nature has fitted me if she has ever fitted any one for anything. Bethink yourself whether you can cast aside all repining and all doubt, and devote yourself in patience and trust to helping me along my path as no one else could. I know what I ask, and the sacrifice I demand, and if this were the time to use false modesty, I should say how little I have to offer in return. . . .

I am full of faults, but I am real and true, and the whole devotion of an earnest soul cannot be over-prized.

. . . It is as if all that old life at Holmwood had merely been a preparation for the real life of our love— as if we were then children ignorant of life's real purpose —as if these last months had merely been my old doubts over again, whether I had rightly or wrongly inter-preted the manner and the words that had given me hope. . . .

We will begin the new love of woman and man, no longer that of boy and girl, conscious that we have aims and purposes as well as affections, and that if love is sweet, life is dreadfully stern and earnest.

As time went on and no permanency offered— although a good deal of writing fell in his way—the strain told heavily upon him. In the autumn he was quite out of sorts, body and mind, more at war with himself than he ever was in his life before. All this, he writes, had darkened his thoughts, had made him once more imagine a hopeless discrepancy be-tween the two of them in their ways of thinking and objects in life. It was not till November 1853 that this depression was banished by the trust and confi-dence of her last letter. "I wish to Heaven," he writes, "it had reached me six months ago. It would have saved me a world of pain and error." But with this, the worst period of mental suffering was over, and every haunting doubt was finally exorcised. His career was made possible by the steady faith which neither separation nor any mis-giving nor its own troubles could shake. And from this point all things began to brighten. His health had been restored by a trip to the Pyrenees with his brother George in September. He had got work that enabled him to regard the Admiralty and its menaces with complete equanimity; a *Manual of Comparative Anatomy*, for Churchill the publisher, regular work on the *Westminster*,[1] and another book in prospect, "so that if I quit the Service to-morrow, these will give me more than my pay has been." And on December 7 he writes how he has been restored and revived by reading over her last two letters, and confesses, "I have been unjust to the depth and strength of your devotion, but will never be so again." Then he tells all he had gone through before leaving England in September for his holiday —how he had resolved to abandon all his special pursuits and take up Chemistry, for practical purposes, when first one publisher and then another asked him to write for them, and hopes were held out to him of being appointed to deliver the Fullerian lectures at the Royal Institution for the next three years; while, most important of all, Edward Forbes was likely, before long, to leave his post at the Museum of Practical Geology, and he had already been spoken

---

[1] This regular work was the article on Contemporary Science, which in October 1854 he got Tyndall to share with him. For, he writes, "To give some account of the books in one's own department is no particular trouble, and comes with me under the head of being paid for what I *must*, in any case, do—but I neither will, nor can, go on writing about books in other departments, of which I am not competent to form a judgment even if I had the time to give to them."

to by the authorities about filling it. This was
worth some £200 a year, while he calculated to make
about £250 by his pen alone. "Therefore it would
be absurd to go hunting for chemical birds in the
bush when I have such in the hand."

CHAPTER VII

1851–1853

SEVERAL letters dating from 1851 to 1853 help to fill up the outlines of Huxley's life during those three years of struggle. There is a description of the British Association meeting at Ipswich in 1851,[1] with the traditional touch of gaiety to enliven the gravity of its proceedings, and the unconventional jollity of the Red Lion Club (a dining-club of members of the Association), whose palmy days were those under the inspiration of the genial and gifted Forbes. This was the meeting at which Huxley first began his alliance with Tyndall, with whom he travelled down from town, although he does not mention his name in this letter. With Hooker he had already made acquaintance; and from this time forwards the three were closely bound together by personal regard as well as by similarity of aims and interests.

[1] "Forbes advises me to go down to the meeting of the British Association this year and make myself notorious somehow or other. Thank Heaven, I have impudence enough to lecture the savans of Europe if necessary. Can you imagine me holding forth?" (June 6, 1851.)

Then follow his sketch of the English scientific world as he found it in 1851, given in his letter to W. Macleay; several letters to his sister; the description of his first lecture at the Royal Institution, which, though successful on the whole, was very different in manner and delivery from the clear and even flow of his later style, with the voice not loud but distinct, the utterance never hurried beyond the point of immediate comprehension, but carrying the attention of the audience with it, eager to the end. Two letters of warning and remonstrance against the habits of lecturing in a colloquial tone, suitable to a knot of students gathered round his table, but not to a large audience,—of running his words, especially technical terms, together,—of pouring out new and unfamiliar matter at breakneck speed, were addressed to him—one by a "working man" of his Monday evening audience at Jermyn Street in 1855, the other, undated, by Mr. Jodrell, a frequenter of the Royal Institution, and afterwards founder of the Jodrell Lectureships at University College, London, and other benefactions to science, and these he kept by him as a perpetual reminder, labelled "Good Advice." How much can be done by the frank acceptance of criticism and by careful practice is shown by the difference between the feelings of the later audiences who flocked to his lectures, and those of the members of an Institute in St. John's Wood, who, as he often used to tell, after hearing him in his early days, petitioned "not to have that young man again."

July 12, 1851.—The interval between my letters has been a little longer than usual, as I have been very busy attending the meeting of the British Association at Ipswich. The last time I attended one was at Southampton five years ago, when I went merely as a spectator, and looked at the people who read papers as if they were somebodies.[1] This time I have been behind the scenes myself and have played out my little part on the boards. I know all about the scenery and decorations, and no longer think the manager a wizard.

Any one who conceives that I went down from any especial interest in the progress of science makes a great mistake. My journey was altogether a matter of policy, partly for the purpose of doing a little necessary trumpeting, and partly to get the assistance of the Association in influencing the Government.

On the journey down, my opposite in the railway carriage turned out to be Sir James Ross, the Antarctic discoverer. We had some very pleasant talk together. I knew all about him, as Dayman[2] had sailed under his command ; oddly enough we afterwards went to lodge at the same house, but as we were attending our respective sections all day we did not see much of one another.

When we arrived at Ipswich there was a good deal of trouble about getting lodgings. My companions located themselves about a mile out of the town, but that was too far for my "indolent habits"; I sought and at last found a room in the town a little bigger than my cabin on board ship, for which I had the satisfaction of paying 30s. a week.

You know what the British Association is. It is a meeting of the savans of England and the Continent under the presidency of some big-wig or other,—this year of the Astronomer-Royal,—for the purpose of

---

[1] See Chap. II., *ad fin.*
[2] One of the lieutenants of the *Rattlesnake.*

exchanging information. To this end they arrange themselves into different sections, each with its own president and committee, and indicated by letters. For instance, Section A is for Mathematics and Physics; Section B for Chemistry, etc.; my own section, that of Natural History, was D, under the presidency of Professor Henslow [1] of Cambridge. I was on the committee, and therefore saw the working of the whole affair.

On the first day there was a dearth of matter in our section. People had not arrived with their papers. So by way of finding out whether I could speak in public or not, I got up and talked to them for about twenty minutes. I was considerably surprised to find that when once I had made the plunge, my tongue went glibly enough.

On the following day I read a long paper, which I had prepared and illustrated with a lot of big diagrams, to an audience of about twenty people! The rest were all away after Prince Albert, who had been unfortunately induced to visit the meeting, and fairly turned the heads of the good people of Ipswich. On Saturday a very pleasant excursion on scientific pretences, but in fact a most jolly and unscientific picnic, took place. Several hundred people went down the Orwell in a steamer. The majority returned, but I and two others, considering Sunday in Ipswich an impossibility, stopped at a little seaside village, Felixstowe, and idled away our time there very pleasantly. Babington [2] the botanist and myself walked in to Ipswich on Sunday night. It is about eleven miles, and we did it comfortably in two hours and three-quarters, which was not bad walking.

---

[1] John Stevens Henslow, 1796-1861, was professor first of Mineralogy, then of Botany, at Cambridge, among his pupils being Charles Darwin. In 1839 he accepted the living of Hitcham in Suffolk, where, by means of a village club, he awakened interest in scientific subjects, especially botany and agriculture.

[2] Charles Cardale Babington, 1808-1895, succeeded Henslow in the Cambridge Chair of Botany, 1861.

On Monday at Section D again. Forbes brought
forward the subject of my application to Government
in committee, and it was unanimously agreed to forward
a resolution on the subject to the Committee of Recommendations. I made a speechification of some length in
the Section about a new animal.

On Thursday morning I attended a meeting of the Ray
Society, and to my infinite astonishment, the secretary,
Dr. Lankester, gave me the second motion to make. The
Prince of Canino moved the first, so I was in good
company. The great absurdity of it was that not being a
member of the Society I had properly no right to speak
at all. However, it was only a vote of thanks, and I got
up and did the " neat and appropriate " in style.

After this a party of us went out dredging in the
Orwell in a small boat. We were away all day, and
it rained hard coming back, so that I got wet through,
and had to pull five miles to keep off my enemy, the
rheumatics.

Then came the President's dinner, to which I did not
go, as I preferred making myself comfortable with a few
friends elsewhere. And after that, the final evening
meeting, when all the final determinations are announced.

Among them I had the satisfaction to hear that it
was resolved—that the President and Council of the
British Association should co-operate with the Royal
Society in representing the value and importance, etc.,
of Mr. T. H. Huxley's zoological researches to Her
Majesty's Government for the purpose of obtaining a
grant towards their publication. Subsequently I was
introduced to Colonel Sabine, the President of the
Association in 1852, and a man of very high standing
and considerable influence. He had previously been
civil enough to sign my certificate at the Royal Society,
unsolicited, and therefore knew me by reputation—I
only mean that as a very small word. He was very
civil and promised me every assistance in his power.

It is a curious thing that out of the four applications
to Government to be made by the Association, two were
for Naval Assistant-Surgeons, viz. one for Dr. Hooker,
who had just returned from the Himalaya Mountains,
and one for me. How I envied Hooker; he has long
been engaged to a daughter of Professor Henslow's, and
at this very meeting he sat by her side. He is going
to be married in a day or two. His father is director
of the Kew Gardens, and there is little doubt of his
succeeding him.

Whether the Government accede to the demand that
will be made upon them or not, I can now rest satisfied
that no means of influencing them has been left unused
by me. If they will not listen to the conjoint recommendations of the Royal Society and the British Association, they will listen to nothing. . . .

July 16, 1851.—I went yesterday to dine with
Colonel Sabine. We had a long discourse about the
prospects and probable means of existence of young
men trying to make their way to an existence in the
scientific world. I took, as indeed what I have seen
has forced me to take, rather the despairing side of
the question, and said that as it seemed to me England
did not afford even the means of existence to young
men who were willing to devote themselves to science.
However, he spoke cheeringly, and advised me by no
means to be hasty, but to wait, and he doubted not
that I should succeed. He cited his own case as an
instance of waiting, eventually successful. Altogether I
felt the better for what he said. . . .

There has been a notice of me in the *Literary Gazette*
for last week, much more laudatory than I deserve, from
the pen of my friend Forbes.[1] . . .

---

[1] An appreciation of his papers on the Physophoridæ and
Sagitta, speaking highly both of his observations and philosophic
power, in the report of the proceedings in Section D.

In the same number is a rich song from the same fertile and versatile pen, which was sung at one of our Red Lion meetings. That is why I want you to look at it, not that you will understand it, because it is full of allusions to occurrences known only in the scientific circles. At Ipswich we had a grand Red Lion meeting; about forty members were present, and among them some of the most distinguished members of the Association. Some foreigners were invited (the Prince of Canino, Buonaparte's nephew, among others), and were not a little astonished to see the grave professors, whose English solemnity and gravity they had doubtless commented on elsewhere, giving themselves up to all sorts of fun. Among the Red Lions we have a custom (instead of cheering) of waving and wagging one coat-tail (one Lion's tail) when we applaud. This seemed to strike the Prince's fancy amazingly, and when he got up to return thanks for his health being drunk, he told us that as he was rather out of practice in speaking English, he would return thanks in our fashion, and therewith he gave three mighty roars and wags, to the no small amusement of every one. He is singularly like the portraits of his uncle, and seems a very jolly, good-humoured old fellow. I believe, however, he is a bit of a rip. It was remarkable how proud the Quakers were of being noticed by him.

TO W. MACLEAY, OF SYDNEY

41 NORTH BANK, REGENT'S PARK,
Nov. 9, 1851.

MY DEAR SIR—It is a year to-day since the old *Rattlesnake* was paid off, and that reminds me among other things that I have hardly kept my promise of giving you information now and then upon the state of matters scientific in England. My last letter is, I am

afraid, nine or ten months old, but here in England the
fighting and scratching to keep your place in the crowd
exclude almost all other thoughts. When I last wrote I
was but at the edge of the crush at the pit-door of
this great fools' theatre—now I have worked my way into
it and through it, and am, I hope, not far from the
check-takers. I have learnt a good deal in my passage.

[Follows an account of his efforts to get his papers
published—substantially a repetition of what has
already been given.]

Rumours there are scattered abroad of a favourable
cast, and I am told on all hands that something will
certainly be done. I only asked for £300, something
less than the cost of a parliamentary blue-book which
nobody ever hears of. They take care to obliterate any
spark of gratitude that might perchance arise for what
they do, by keeping one so long in suspense that the
result becomes almost a matter of indifference. Had I
known they would keep me so long, I would have
published my work as a series of papers in the *Philosophical Transactions.*

In the meanwhile I have not been idle, as I hope to
show you by the various papers enclosed with this. You
will recollect that on the Salpæ. No one here knew
anything about them, and I thought that all my results
were absolutely new—until, *me miserum!* I found them
in a little paper of Krohn's in the *Annales des Sciences* for
1846, without any figures to draw anybody's attention.

The memoir on the Medusæ (which I sent to you) has,
I hear, just escaped a high honour—to wit, the Royal
Medal. The award has been made to Newport for his
paper on "Impregnation." I had no idea that anything
I had done was likely to have the slightest claim to such
distinction, but I was informed yesterday by one of the
Council that the balance hung pretty evenly, and was

only decided by their thinking my memoir was too small and short.

I have been working in all things with a reference to wide views of zoological philosophy, and the report upon the Echinoderms is intended in common with the mem. on the Salpæ to explain my views of Individuality among the lower animals—views which I mean to illustrate still further and enunciate still more clearly in my book that is to be.[1] They have met with approval from Carpenter, as you will see by the last edition of his *Principles of Physiology*, and I think that Forbes and some others will be very likely eventually to come round to them, but everything that relates to abstract thought is at a low ebb among the mass of naturalists in this country.

In the paper upon " Thalassicolla " and in that which I read before the British Association, as also in one upon the organisation of the Rotifera, which I am going to have published in the Microscopical Society's *Transactions*, I have been driving in a series of wedges into Cuvier's Radiata, and showing how *selon moi* they ought to be distributed.

I am every day becoming more and more certain that you were on the right track thirty years ago in your views of the order and symmetry to be traced in the true natural system.

During the next session I mean to send in a paper to the R.S. upon the " Homologies of the Mollusca," which shall astonish them. I want to get done for the Mollusca what Savigny did for the Articulata, viz. to show how they all—Cephalopoda, Gasteropoda, Pteropoda, Heteropoda, etc.—are organised on one type, and how the homologous organs are modified in each. What with this and the book, I shall have enough to do for the next six months.

You will doubtless ask what is the practical outlook

---

[1] He lectured on this subject at the Royal Institution in 1852.

of all this? whether it leads anywhere in the direction
of bread and cheese? To this also I can give a tolerably
satisfactory answer.

As you *won't* have a Professor of Natural History at
Sydney—to my great sorrow—I have gone in as a
candidate for a Professorial chair at the other end of the
world, Toronto in Canada. In England there is nothing
to be done—it is the most hopeless prospect I know of;
of course the Service offers nothing for me except irre-
trievable waste of time, and the scientific appointments
are so few and so poor that they are not tempting. . . .

Had the Sydney University been carried out as
originally proposed, I should certainly have become a
candidate for the Natural History Chair. I know no
finer field for exertion for any naturalist than Sydney
Harbour itself. Should such a Professorship be hereafter
established, I trust you will jog the memory of my
Australian friends in my behalf. I have finally decided
that my vocation is science, and I have made up my mind
to the comparative poverty which is its necessary adjunct,
and to the no less certain seclusion from the ordinary
pleasures and rewards of men. I say this without the
slightest idea that there is anything to be enthusiastic
about in either science or its professors. A year behind
the scenes is quite enough to disabuse one of all rose-pink
illusions.

But it is equally clear to me that for a man of my
temperament, at any rate, the sole secret of getting
through this life with anything like contentment is to
have full scope for the development of one's faculties.
Science alone seems to me to afford this scope—Law,
Divinity, Physic, and Politics being in a state of chaotic
vibration between utter humbug and utter scepticism.

There is a great stir in the scientific world at present
about who is to occupy Konig's place at the British
Museum, and whether the whole establishment had better
not, *quoad* Zoology, be remodelled and placed under

Owen's superintendence. The heart-burnings and jealousies about this matter are beyond all conception. Owen is both feared and hated, and it is predicted that if Gray and he come to be officers of the same institution, in a year or two the total result will be a caudal vertebra of each remaining after the manner of the Kilkenny cats.

However, I heard yesterday, upon what professed to be very good authority, that Owen would not leave the College under any circumstances.

It is astonishing with what an intense feeling of hatred Owen is regarded by the majority of his contemporaries, with Mantell as arch-hater. The truth is, he is the superior of most, and does not conceal that he knows it, and it must be confessed that he does some very ill-natured tricks now and then. A striking specimen of one is to be found in his article on Lyell in the last *Quarterly*, where he pillories poor Quekett—a most inoffensive man and his own immediate subordinate —in a manner not more remarkable for its severity than for its bad taste.[1] That review has done him much harm in the estimation of thinking men—and curiously enough, since it was written, reptiles have been found in the old red sandstone, and insectivorous mammals in the Trias ! Owen is an able man, but to my mind not so great as he thinks himself. He can only work in the concrete from bone to bone, in abstract reasoning he becomes lost— witness "Parthenogenesis" which he told me he considered one of the best things he had done !

He has, however, been very civil to me, and I am as grateful as it is possible to be towards a man with whom I feel it necessary to be always on my guard.

Quite another being is the other leader of Zoological Science in this country—I mean Edward Forbes, Paleontologist to the Geological Survey. More especially

---

[1] Cp. p. 242, note *ad fin.*

a Zoologist and a Geologist than a Comparative Anatomist, he has more claims to the title of a Philosophic Naturalist than any man I know of in England. A man of letters and an artist, he has not merged the *man* in the man of science—he has sympathies for all, and an earnest, truth-seeking, thoroughly genial disposition which win for him your affection as well as your respect. Forbes has more influence by his personal weight and example upon the rising generation of scientific naturalists than Owen will have if he write from now till Doomsday.

Personally I am greatly indebted to him (though the opinion I have just expressed is that of the world in general). During my absence he superintended the publication of my paper, and from the moment of my arrival until now he has given me all the help one man can give another. Why he should have done so I do not know, as when I left England I had only spoken to him once.

The rest of the naturalists stand far below these two in learning, originality, and grasp of mind. Goodsir of Edinburgh should, I. suppose, come next, but he can't write intelligibly. Darwin might be anything if he had good health. Bell is a good man in all the senses of the word, but wants qualities 2 and 3. Newport a laborious man, but wants 1 and 3. Grant and Rymer Jones— *arcades ambo*—have mistaken their vocation.

My old chief Richardson is a man of men, but troubles himself little with anything but detail zoology. What think you of his getting married for the third time just before his last expedition. I hardly know by which step he approved himself the bolder man.

I think I have now fulfilled my promise of supplying you with a little scientific scandal—and if this long epistle has repaid your trouble in getting through it, I am content.

Believe me, I have not forgotten, nor ever shall forget, your kindness to me at a time when a little appreciation and encouragement were more grateful to me and of more service than they will perhaps ever be again. I have done my best to justify you.

I send copies of all the papers I have published with one exception, of which I have none separate. Of the Royal Society papers I send a double set. Will you be kind enough to give one with my kind regards and remembrances to Dr. Nicholson? I feel I ought to have written to him before leaving Sydney, but I trust he will excuse my not having done so.

I shall be very glad if you can find time to write.—
Ever yours faithfully,   T. H. HUXLEY.

W. Macleay, Esq.

P.S.—Müller has just made a most extraordinary discovery, no less than the generation of Molluscs from Holothuriæ ! ! ! You will find a translation of his paper by me in the *Annals* for January 1852.

Dec. 13, 1851.

TO HIS SISTER

May 20, 1851.

. . . Owen has been amazingly civil to me, and it was through his writing to the First Lord that I got my present appointment. He is a queer fish, more odd in appearance than ever . . . and more bland in manner. He is so frightfully polite that I never feel thoroughly at home with him. He got me to furnish him with some notes for the second edition of the *Admiralty Manual of Scientific Inquiry*, and I find that in it Darwin and I (comparisons are odorous) figure as joint authorities on some microscopic matters ! !

Professor Forbes, however, is my great ally, a first-rate man, thoroughly in earnest and disinterested, and ready to give his time and influence—which is great—to help any man who is working for the cause. To him

I am indebted for the supervision of papers that were published in my absence, for many introductions, and most valuable information and assistance, and all done in such a way as not to oppress one or give one any feeling of patronage, which you know (so much do I retain of my old self) would not suit me. My notions are diametrically opposed to his in some matters, and he helps me to oppose him. The other night, or rather nights, for it took three, I had a long paper read at the Royal Society which opposed some of his views, and he got up and spoke in the highest terms of it afterwards. This is all as it should be. I can reverence such a man and yet respect myself.

I have been aspiring to great honours since I wrote to you last, to wit the F.R.S., and found no little to my astonishment that I had a chance of it, and so went in. I must tell you that they have made the admission more difficult than it used to be. Candidates are not elected by the Society alone, but fifteen only a year are selected by a committee, and then elected as a matter of course by the Society. This year there were thirty-eight candidates. I did not expect to come in till next year, but I find I am one of the selected. I fancy I shall be the junior Fellow by some years. Singularly enough, among the non-selected candidates were Ward, the man who conducted the Botanical Honours Examination of Apothecaries' Hall nine years ago, and Bryson, the surgeon of the *Fisguard*, *i.e.*, nominally my immediate superior, and who, as he frequently acts as Sir Wm. Burnett's deputy, *will very likely examine me when I pass for Surgeon R.N.! !* That is awkward and must be annoying to him, but it is not my fault. I did not ask for a single name that appeared upon my certificate. Owen's name and Carpenter's, which were to have been appended, were not added. Forbes, my recommender, told me beforehand not to expect to get in this year, and did not use his influence, and so I have no intriguing to reproach myself with or to be reproached with. The only drawback is that it will cost me £14, which is more than I can very well afford.

By the way, I have not told you that after staying for about five months with George, I found that if I meant to work in earnest his home was not the place, so, much to my regret,—for they made me very happy there,—I summoned resolution and *The Boy's Own Book* and took a den of my own, whence I write at present. You had better, however, direct to George, as I am going to move and don't know how long I may remain at my next habitation. At present I am living in the Park Road, but I find it too noisy and am going to St. Anne's Gardens, St. John's Wood, close to my mother's, against whose forays I shall have to fortify myself.

It was a minor addition to his many troubles that after a time Huxley found a grudging and jealous spirit exhibited in some quarters towards his success, and influence used to prevent any further advance that might endanger the existing balance of power in the scientific world. But this could be battled with directly; indeed it was rather a relief to have an opportunity for action instead of sitting still to wait the results of uncertain elections. The qualities requisite for such a contest he possessed, in a high ideal of the dignity of science as an instrument of truth; a standard of veracity in scientific workers to which all should subordinate their personal ambitions; a disregard of authority as such unless its claims were verified by indisputable fact; and as a beginning, the will to subject himself to his own most rigid canons of accuracy, thorough-

ness, and honesty ; then to maintain his principle
and defend his position against all attempts at
browbeating.

March 5, 1852.

I told you I was very busy, and I must tell you what
I am about and you will believe me.   I have just finished
a Memoir for the Royal Society,[1] which has taken me a
world of time, thought, and reading, and is, perhaps, the
best thing I have done yet.   It will not be read till May,
and I do not know whether they will print it or not
afterwards ; that will require care and a little manœuvring
on my part.   You have no notion of the intrigues that
go on in this blessed world of science.   Science is, I fear,
no purer than any other region of human activity ; though
it should be.   Merit alone is very little good ; it must
be backed by tact and knowledge of the world to do very
much.

For instance, I know that the paper I have just sent
in is very original and of some importance, and I am
equally sure that if it is referred to the judgment of my
"particular friend " —— that it will not be published.
He won't be able to say a word against it, but he will
pooh-pooh it to a dead certainty.

You will ask with some wonderment, Why ?   Because
for the last twenty years —— has been regarded as the
great authority on these matters, and has had no one to
tread on his heels, until at last, I think, he has come to
look upon the Natural World as his special preserve, and
"no poachers allowed."   So I must manœuvre a little to
get my poor memoir kept out of his hands.

The necessity for these little stratagems utterly disgusts
me.   I would so willingly reverence and trust any man
of high standing and ability.   I am so utterly unable to

---

[1] "On the Morphology of the Cephalous Mollusca," *Scientific
Memoirs,* vol. i. p. 152.

comprehend this petty greediness. And yet withal you will smile at my perversity. I have a certain pleasure in overcoming these obstacles, and fighting these folks with their own weapons. I do so long to be able to trust men implicitly. I have such a horror of all this literary pettifogging. I could be so content myself, if the necessity of making a position would allow it, to work on anonymously, but —— I see is determined not to let either me or any one else rise if he can help it. Let him beware. On my own subjects I am his master, and am quite ready to fight half a dozen dragons. And although he has a bitter pen, I flatter myself that on occasions I can match him in that department also.

But I was telling you how busy I am. I am getting a memoir ready for the Zoological Society, and working at my lecture for the Royal Institution, which I want to make striking and original, as it is a good opportunity, besides doing a translation now and then for one of the Journals. Besides this, I am working at the British Museum to make a catalogue of some creatures there. All these things take a world of time and labour, and yield next to no direct profit; but they bring me into contact with all sorts of men, in a very independent position, and I am told, and indeed hope, that something must arise from it. So fair a prospect opens out before me if I can only wait. I am beginning to know what *work* means, and see how much more may be done by steady, unceasing, and well-directed efforts. I thrive upon it too. I am as well as ever I was in my life, and the more I work the better my temper seems to be.

April 30, 1852, 11½ P.M.

I have just returned from giving my lecture[1] at the Royal Institution, of which I told you in my last letter.

---

[1] "On Animal Individuality," *Scientific Memoirs*, vol. i. p. 146, cp. p. 134, *supra.*

I had got very nervous about it, and my poor mother's death had greatly upset my plans for working it out.

It was the first lecture I had ever given in my life, and to what is considered the best audience in London. As nothing ever works up my energies but a high flight, I had chosen a very difficult abstract point, in my view of which I stand almost alone. When I took a glimpse into the theatre and saw it full of faces, I did feel most amazingly uncomfortable. I can now quite understand what it is to be going to be hanged, and nothing but the necessity of the case prevented me from running away.

However, when the hour struck, in I marched, and began to deliver my discourse. For ten minutes I did not quite know where I was, but by degrees I got used to it, and gradually gained perfect command of myself and of my subject. I believe I contrived to interest my audience, and upon the whole I think I may say that this essay was successful.

Thank Heaven I can say so, for though it is no great matter succeeding, failing would have been a bitter annoyance to me. It has put me comfortably at my ease with regard to all future lecturings. After the Royal Institution there is no audience I shall ever fear.

May 9.

The foolish state of excitement into which I allowed myself to get the other day completely did for me, and I have hardly done anything since except sleep a great deal. It is a strange thing that with all my will I cannot control my physical organisation.

To his Sister

April 17, 1852.

. . . I fear nothing will have prepared you to hear that one so active in body and mind as our poor mother was has been taken from us. But so it is. . . .

It was very strange that before leaving London my mother, possessed by a strange whim, as I thought, distributed to many of us little things belonging to her. I laughed at her for what I called her "testamentary disposition," little dreaming that the words were prophetic.

[The summons to those of the family in London reached them late, and their arrival was made still later by inconvenient trains and a midnight drive, so that all had long been over when they came to Barning in Kent, where the elder Huxleys had just settled near their son James.]

Our mother had died at half-past four, falling gradually into a more and more profound insensibility. She was thus happily spared the pain of fruitlessly wishing us round her, in her last moments; and as the hand of Death was upon her, I know not that it could have fallen more lightly.

I offer you no consolation, my dearest sister, for I know of none. There are things which each must bear as he best may with the strength that has been allotted to him. Would that I were near you to soften the blow by the sympathy which we should have in common. . . .

May 3, 1852.

So much occupation has crowded upon me between the beginning of this letter and the present time that I have been unable to finish it. I had undertaken to give a lecture at the Royal Institution on the 30th April. It was on a difficult subject, requiring a good deal of thought ; and as it was my first appearance and before the best audience in London, you may imagine how anxious and nervous I was, and how completely I was obliged to abstract my thoughts from everything else.

However, I am happy to say it is well over. There was a very good audience—Faraday, Prof. Forbes, Dr. Forbes, Wharton Jones, and [a] whole 'lot of "nobs," among my auditors. I had made up my mind all day to break down, and then go and hang myself privately. And so you may imagine that I entered the theatre with a very pale face, and a heart beating like a sledge-hammer nineteen to the dozen. For the first five minutes I did not know very clearly what I was about, but by degrees I got possession of myself and of my subject, and did not care for anybody. I have had "golden opinions from all sorts of men" about it, so I suppose I may tell you I have succeeded. I don't think, however, that I ever felt so thoroughly used up in my life as I did for two days afterwards. There is one comfort, I shall never be nervous again about any audience; but at one's first attempt, to stand in the place of Faraday and such big-wigs might excuse a little weakness.

The way is clear before me, if my external circumstances will only allow me to persevere; but I fully expect that I shall have to give up my dreams.

Science in England does everything—but *pay*. You may earn praise but not pudding.

I have helping hands held out to me on all sides, but there is nothing to help me to. Last year I became a candidate for a Professorship at Toronto. I took an infinity of trouble over the thing, and got together a mass of testimonials and recommendations, much better than I had any right to expect. From that time to this I have heard nothing of the business—a result for which I care the less, as I believe the chair will be given to a brother of one of the members of the Canadian ministry, who is, I hear, a candidate. Such a qualification as that is, of course, better than all the testimonials in the world.

I think I told you when I last wrote that I was expecting a grant from Government to publish the chief part of my work, done while away. I am expecting it still. I got tired of waiting the other day and wrote to the Duke of Northumberland, who is at present First Lord of the Admiralty, upon the subject. His Grace has taken the matter up, and I hope now to get it done.

With all this, however, Time runs on. People look upon me, I suppose, as a "very promising young man," and perhaps envy my "success," and I all the while am cursing my stars that my Pegasus *will* fly aloft instead of pulling slowly along in some respectable gig, and getting his oats like any other praiseworthy cart-horse.

It's a charming piece of irony altogether. It is two years yesterday since I left Sydney harbour—and of course as long since I saw Nettie. I am getting thoroughly tired of our separation, and I think she is, though the dear little soul is ready to do anything for my sake, and yet I dare not face the stagnation—the sense of having failed in the whole purpose of my existence—which would, I know, sooner or later beset me, even with her, if I forsake my present object. Can you wonder with all this, my dearest Lizzie, that often as I long for your brave heart and clear head to support and advise me, I yet rarely feel inclined to write? Pray write to me more often than you have done; tell me all about yourself and the Doctor and your children. They must be growing up fast, and Florry must be getting beyond the "Bird of Paradise" I promised her. Love and kisses to all of them, and kindest remembrances to the Doctor.— Ever your affectionate brother,   T. H. HUXLEY.

To Miss HEATHORN

Nov. 13, 1852.

Going last week to the Royal Society's library for a book, and like the boy in church "thinkin' o' naughten," when I went in, Weld, the Assistant Secretary, said, "Well, I congratulate you." I confess I did not see at that moment what any mortal man had to congratulate me about. I had a deuced bad cold, with rheumatism in my head; it was a beastly November day and I was very grumpy, so I inquired in a state of mild surprise

what might be the matter. Whereupon I learnt that
the Medal had been conferred at the meeting of the
Council on the day before. I was very pleased . . . and
I thought you would be so too, and I thought moreover
that it was a fine lever to help us on, and if I could
have sent a letter to you immediately I should have sat
down and have written one to you on the spot. As it is
I have waited for official confirmation and a convenient
season.

And now . . . shall I be very naughty and make a
confession ? The thing that a fortnight ago (before I got
it) I thought so much of, I give you my word I do not
care a pin for. I am sick of it and ashamed of having
thought so much of it, and the congratulations I get give
me a sort of internal sardonic grin. I think this has
come about partly because I did not get the official confirmation of what I had heard for some days, and with my
habit of facing the ill side of things I came to the conclusion that Weld had made a mistake, and I went in
thought through the whole enormous mortification of
having to explain to those to whom I had mentioned it
that it was quite a mistake. I found that all this, when
I came to look at it, was by no means so dreadful as it
seemed—quite bearable in short—and then I laughed at
myself and have cared nothing about the whole concern
ever since. In truth . . . I do not think that I am in
the proper sense of the word ambitious. I have an
enormous longing after the highest and best in all shapes—
a longing which haunts me and is the demon which ever
impels me to work, and will let me have no rest unless I
am doing his behests. The honours of men I value so far
as they are evidences of power, but with the cynical mis-
trust of their judgment and my own worthiness, which
always haunts me, I put very little faith in them. Their
praise makes me sneer inwardly. God forgive me if I do
them any great wrong.

. . . I feel and know that all the rewards and honours

in the world will ever be worthless for me as soon as they are obtained. I know that always, as now, they will make me more sad than joyful. I know that nothing that could be done would give me the pure and heartfelt joy and peace of mind that your love has given me, and, please God, shall give for many a long year to come, and yet my demon says work ! work ! you shall not even love unless you work.

Not blinded by any vanity, then, I hope . . . but viewing this stroke of fortune as respects its public estimation only, I think I must look upon the award of this medal as the turning-point of my life, as the finger-post teaching me as clearly as anything can what is the true career that lies open before me. For whatever may be my own private estimation of it, there can be no doubt as to the general feeling about this thing, and in case of my candidature for any office it would have the very greatest weight. And as you will have seen by my last letter, it only strengthens and confirms the conclusion I had come to. Bid me God-speed then . . . it is all I want to labour cheerfully.

Nov. 28.

. . . You will hear all the details of the Great Duke's state funeral from the papers much better than I can tell you them. I went to the Cathedral (St. Paul's) and had the good fortune to get a capital seat—in front, close to the great door by which every one entered. It was bitter cold, a keen November wind blowing right in, and as I was there from eight till three, I expected nothing less than rheumatic fever the next day ; however I didn't get it. It was pitiful to see the poor old Marquis of Anglesey—a year older than the Duke—standing with bare head in the keen wind close to me for more than three-quarters of an hour. It was impressive enough—the great interior lighted up by a single line of light running along the whole circuit of the cornice, and another encircling the

dome, and casting a curious illumination over the masses
of uniforms which filled the great space. The best of our
people were there and passed close to me, but the only
face that made any great impression upon my memory
was that of Sir Chas. Napier, the conqueror of Scinde.
Fancy a very large, broad-winged, and fierce-looking
hawk in uniform. Such an eye !

When the coffin and the mourners had passed I closed
up with the soldiers and went up under the dome, where
I heard the magnificent service in full perfection.

All of it, however, was but stage trickery compared
with the noble simplicity of the old man's life. How the
old stoic, used to his iron bed and hard hair pillow, would
have smiled at all the pomp—submitting to that, however,
and all other things necessary to the "carrying on of the
Queen's Government."

I send Tennyson's ode by way of packing—it is not
worth much more, the only decent passages, to my mind,
being those I have marked.

The day after to-morrow I go to have my medal presented and to dine and make a speech.

The Royal Medal was conferred on November 30,
and the medallists were entertained at the anniversary
dinner of the Society on that day.   In the words
with which the President, the Earl of Rosse, accompanied the presentation of the medal, "it is not
difficult," writes Sir M. Foster, "reading between
the lines, to recognise the appreciation of a new
spirit of anatomical inquiry, not wholly free from a
timorous apprehension as to its complete validity."[1]

[1] "In these papers (on the Medusæ) you have for the first time
fully developed their structure, and laid the foundation of a
rational theory for their classification." "In your second paper
'On the Anatomy of Salpa and Pyrosoma,' the phenomena, etc.,
have received the most ingenious and elaborate elucidation, and

For the difference between this and the labours of
the greatest English comparative anatomist of the
time, whose detailed work was of the highest value,
but whose generalisations and speculations, based on
the philosophy of Oken, proved barren and fruitless,
lay in the fact that Huxley, led to it doubtless by
his solitary readings in his Charing Cross days, had
taken up the method of von Baer and Johannes
Müller, then almost unknown, or at least unused in
England—"the method which led the anatomist to
face his problems in the spirit in which the physicist
faced his."

He had been warned by Forbes not to speak too
strongly about the dilatoriness of the Government
in the matter of the grant, so he writes: "I will
'roar you like any sucking dove' at the dinner,
though I felt tempted otherwise." On December 1
he tells how he carried out this advice.

MY DEAR FORBES—You will, I know, like to learn
how I got on yesterday. The President's address to me
had been drawn up by Bell. It was, of course, too
flattering, but he had taken hold of the right points in
my work—at least I thought so.

Bunsen spoke very well for Humboldt.

There was a capital congregation at the dinner—sixty
or seventy Fellows there. . . .

When it came to my turn to return thanks, I believe

have given rise to a process of reasoning, the results of which can
scarcely yet be anticipated, but must bear in a very important
degree upon some of the most abstruse points of what may be
called transcendental physiology." See *Royal Society*, Obituary
Notices, vol. lix. p. 1.

I made a very tolerable speechification, at least everybody says so. Lord Rosse had alluded to "science having to take care of itself in this country," and in winding up I gave them a small screed upon that text. That you may see I kept your caution in mind, I will tell you as nearly as may be what I said. I told them that I could not conceive that anything I had hitherto done merited the honour of that day (I looked so preciously meek over this), but that I was glad to be able to say that I had so much unpublished material as to make me hopeful of one day diminishing the debt. I then said, "The Government of this country, of this *great* country, has been two years debating whether it should grant the three hundred pounds necessary for the publication of these researches. I have been too long used to strict discipline to venture to criticise any act of my superiors, but I venture to hope that before long, in consequence of the exertions of Lord Rosse, of the President of the British Association, and the goodwill, which I gratefully acknowledge, of the present Lord of the Admiralty, I shall be able to lay before you something more worthy of to-day's award."

I had my doubts how the nobs would take it, but both Lord Rosse and Sabine warmly commended my speech and regretted I had not said even more upon the subject.

Some light is thrown upon his habits at this time by the following, part of his letter to Forbes of November 19 :—

I have frequent visits from ——. He is a good man, but direfully argumentative, and in that sense to me a bore. Besides that, the creature will come and call upon me at nine or ten o'clock in the morning before I am out of bed, or if out of bed, before I am in possession of my faculties, which never arrive before twelve or one.

This morning incapacity was of a piece with his hatred of the breakfast-party of the period. To go abroad from home or to do any work before break-fasting ensured him a headache for the rest of the day, so that he never was one of those risers with the dawn who do half a day's work before the rest of the world is astir. And though necessity often compelled him to do with less, he always found eight hours his proper allowance of sleep.

But in the end of 1853 we hear of a reform in his ways, after a bad bout of ill-health, when he rises at eight, goes to bed at twelve, and eschews parties of every kind as far as possible, with excellent results as far as health went.

After his marriage, however, and indeed to the beginning of his last illness, he always rose early enough for an eight o'clock breakfast, after which the working day began, lasting regularly from a little after nine till midnight.

> 4 UPPER YORK PLACE, ST. JOHN'S WOOD,
> *Feb.* 6, 1853.

Many thanks, my dearest sister, for your kind and thoughtful letter—it went to my heart no little that you, amidst all your trials and troubles, should find time to think so wisely and so affectionately of mine. Though greatly tempted otherwise, I have acted in the spirit of your advice, and my reward, in the shape of honours at any rate, has not failed me, as the Royal Society gave me one of the Royal medals last year. It's a bigger one than I got under your auspices so many years ago, being worth £50, but I don't know that I cared so much about it.

It was assigned to me quite unexpectedly, and in the eyes of the world I, of course, am greatly the bigger—but I will confess to you privately that I am by no means dilated, and am the identical Boy Tom I was before I achieved the attainment of my golden porter's badge. Curiously it was given for the first Memoir I have in the Royal Society's *Transactions*, sent home four years ago with no small fear and trembling, and, "after many days," returning with this queer crust of bread. In the speech I had to make at the Anniversary Dinner I grew quite eloquent on that point, and talked of the dove I had sent from my ark, returning, not with the olive branch, but with a sprig of the bay and a fruit from the garden of the Hesperides—a simile which I thought decidedly clever, but which the audience—distinguished audience I ought to have said—probably didn't, as they did not applaud that, while they did some things I said which were incomparably more stupid. This was in November, and I ought to have written to you about it before, my dear Lizzie, but for one thing I am very much occupied, and for the other (shall I confess it ?) I was rather puzzled that I had not heard from you since I wrote. Now my useless conscience, which never makes me do anything right in time, is pitching in to me when it is too late.

The medal, however, must not be jested at, as it is most decidedly of practical use in giving me a status in the eyes of those charming people, "practical men," such as I had not before, and I am amused to find some of my friends, whose contempt for my "dreamy" notions was not small in time past, absolutely advising me to take a far more dreamy course than I dare venture upon. However, I take very much my own course now, even as I have done before—Huxley all over.

However, that is enough about myself just now. In the next letter I will tell you more at length about my plans and prospects, which are mostly, I am sorry to say, only provocative of setting my teeth hard and saying, "Never mind, I *will*." But what I write in a hurry about and want you to do at once, is to write to me and tell me exactly how money may be sent safely to you. It is inexpedient to send without definite directions, according to the character you give your neighbours. Don't expect anything vast, but there is corn in Egypt. . . .

Two classes of people can I deal with and no third. They are the good people —people after my own heart, and the thorough men of the world. Either of these I can act and sympathise with, but the others, who are neither for God nor for the Devil, but for themselves, as grim old Dante has it, and whom he therefore very justly puts in a most uncomfortable place, I cannot do with. . . .

So Florry is growing up into a great girl ; the child will not remember me, but kiss her and my godson for me, and give my love to them all. The Lymph shall come in my next letter for the young Yankee. I hope the juices of the English cow will prevent him from ever acquiring the snuffle.

Tell the Doctor all about the medal, with my kindest regards, and believe me, my dearest Lizzie, your affectionate brother, TOM.

4 UPPER YORK PLACE, ST. JOHN'S WOOD,
April 22, 1853.

MY DEAREST LIZZIE—First let me congratulate you on being safe over your troubles and in possession of another possible President. I think it may be worth coming over twenty years hence on the possibility of picking up something or other from one of my nephews at Washington.

[He sends some money.] Would it were more worth your having, but I have not as yet got on to Tom Tiddler's ground on this side of the water. You need not be alarmed about my having involved myself in any way—such portion of it as is of my sending has been conquered by mine own sword and spear, and the rest came from Mary.[1]. . .

[After giving a summary of his struggle with the Admiralty, he proceeds]—If I were to tell you all the intriguing and humbug there has been about my unfortunate grant—which is not yet granted—it would occupy this letter, and though a very good illustration of the encouragement afforded to Science in this country, would not be very amusing. Once or twice it has fairly died out, only to be stirred up again by my own pertinacity. However, I have hopes of it at last, as I hear Lord Rosse is just about to make another application to the present Government on the subject. While this business has been dragging on of course I have not been idle. I have four memoirs (on various matters in Comparative Anatomy) in the *Philosophical Transactions,* and they have given me their Fellowship and one of the Royal medals. I have written a whole lot of things for the journals—reviews for the *British and Foreign Quarterly Medical,* etc. I am one of the editors of Taylor's *Scientific Memoirs* (German scientific translations). In conjunction with my friend Busk I am translating a great German book on the *Microscopical Anatomy of Man,* and I have engaged to write a long article for Todd's [2] *Cyclopædia.* Besides this, have read two long memoirs at the British Association, and have given two lectures at the Royal Institution—one of them only two days ago, when I was so ill with influenza I could hardly stand or speak.

Furthermore, I have been a candidate for a Professorship of Natural History at Toronto (which is not even

---

[1] Mrs. George Huxley.
[2] Robert Bentley Todd, 1809-1860, a popular practitioner and professor of Anatomy and Physiology at King's College, London, 1836-53. His *Cyclopædia of Anatomy and Physiology* was published between 1835 and 1859.

yet decided) ; for one at Aberdeen, which has been given against me ; and at present I am a candidate for the Professorship of Physiology at King's College, or, rather, for half of it—Todd having given up, and Bowman,[1] who remains, being willing to take only half, and that he will soon give up. My friend Edward Forbes—a regular brick, who has backed me through thick and thin—is backing me for King's College, where he is one of the Professors. My chance is, I believe, very good, but nothing can be more uncertain than the result of the contest. If they don't take one of their own men, I think they will have me. It would suit me very well, and the whole chair is worth £400 a year, and would enable me to live.

Something I must make up my mind to do, and that speedily. I can get honour in Science, but it doesn't pay, and "honour heals no wounds." In truth I am often very weary. The longer one lives the more the ideal and the purpose vanishes out of one's life, and I begin to doubt whether I have done wisely in giving vent to the cherished tendency towards Science which has haunted me ever since my childhood. Had I given myself to Mammon I might have been a respectable member of society with large watch-seals by this time. I think it is very likely that if this King's College business goes against me, I may give up the farce altogether—burn my books, bury my rod, and take to practice in Australia. It is no use to go on kicking against the pricks. . . .

---

[1] Sir William Bowman, 1816-1892, the ophthalmic surgeon, closely connected with King's College Hospital. He was associated with Todd in preparing his *Cyclopædia* and his *Anatomy of Man.*

CHAPTER VIII

1854

THE year 1854 marks the turning-point in Huxley's career. The desperate time of waiting came to an end. By the help of his lectures and his pen, he could at all events stand and wait independently of the Navy. He could not, of course, think of immediate marriage, nor of asking Miss Heathorn to join him in England; but it so happened that her father was already thinking of returning home, and finally this was determined upon just before Professor Forbes' translation to a chair at Edinburgh gave Huxley what turned out to be the long-hoped-for permanency in London.

June 3, 1854.

I have often spoken to you of my friend Edward Forbes. He has quite recently been suddenly appointed to a Professorial Chair in Edinburgh, vacated by the death of old Jamieson. He was obliged to go down there at once and lecture, and as he had just commenced his course at the Government School of Mines in Jermyn Street, it was necessary to obtain a substitute. He had spoken to me of the possibility of his being called away long ago, and had asked if I would take his place, to which, of course, I assented, but the whole affair was so uncertain that I never in any way reckoned upon it. Even at last I did not know on the Monday whether I was to go on for him on the Friday or not. However, he did go after giving two lectures, and on Friday the 25th May I took his lecture, and I have been going on ever since, twice a week on Mondays and Fridays. Called upon so very suddenly to give a course of some six-and-twenty lectures, I find it very hard work, but I like it and I never was in better health.

On July 20, this temporary work, which he had undertaken as the friend of Forbes, was exchanged for one of the permanent lectureships formerly held by the latter. A hundred a year for twenty-six lectures was not affluence; it would have suited him better to have had twice the work and twice the pay. But it was his crossing of the Rubicon, and, strangely enough, no sooner had he gained this success than it was doubled.

July 30, 1854.

I was appointed yesterday to a post of £200 a year. It has all come about in the strangest way. I told you how my friend Forbes had been suddenly called away to Edinburgh, and that I had suddenly taken his duties— sharp work it has been I can tell you these summer months, but it is over and done satisfactorily. Forbes got £500 a year, £200 for a double lectureship, £300 for another office. I took one of the lectureships, which would have given me £100 a year only, and another man was to have the second lectureship and the other office in question. It was so completely settled a week

ago that I had written to the President of the Board
of Trade who makes the appointment, accepting mine,
and the other man had done the same.   Happily for me,
however, my new colleague was suddenly afflicted with a
sort of moral colic, an absurd idea that he could not
perform the duties of his office, and resigned it.   The
result is that a new man has been appointed to the office
he left vacant, while the lectureship was offered to me.
Of course I took it, and so in the course of the week I
have seen my paid income doubled. . . . So after a short
interval I have become a Government officer again, but
in rather a different position, I flatter myself.   I am
chief of my own department, and my position is considered
a very good one—as good as anything of its kind in
London.

Furthermore, on August 11 he was "entrusted
with the Coast Survey investigations under the
Geological Survey, and remunerated by fee until
March 31, 1855, when he was ranked as Naturalist
on the Survey with an additional salary of £200,
afterwards increased to £400, rising to £600 per
annum," as the official statement has it.

Then in quick succession he was offered in August
a lectureship on Comparative Anatomy at St. Thomas'
Hospital for the following May and June, and in
September he was asked to lecture in November and
March for the Science and Art Department at
Marlborough House.

Now therefore, with the Heathorns coming to
England, his plans and theirs exactly fitted, and he
proposed to get married as soon as they came over,
early in the following summer.

A letter of this year deserves quoting as illustrating the directness of Huxley's dealings with his friends, and his hatred of doing anything unknown to them which might be misreported to them or misconstrued without explanation. As a member of the Royal Society Council, it was his duty to vote upon the persons to whom the yearly medals of the Society should be awarded. For the Royal Medal first Hooker was named, and received his hearty support; then Forbes, in opposition to Hooker, in his eyes equally deserving of recognition, and almost more closely bound to him by ties of friendship, so that whatever action he took, might be ascribed to motives which should have no part in such a selection. The course actually taken by him he explained at length in letters to both Forbes and Hooker.

Nov. 6, 1854.

MY DEAR HOOKER—I have been so busy with lecturing here and there that I have not had time to write and congratulate you on the award of the medal. The queer position in which I was placed prevents me from being able to congratulate *myself* on having any finger in the pie, but I am quite sure there was no member of the Council who felt more strongly than myself that what honour the bauble could confer was most fully won, and no more than your just deserts ; or who rejoiced more when the thing was settled in your favour.

However, I do trust that I shall never be placed in such an awkward position again. I would have given a great deal to be able to back Forbes tooth and nail—not only on account of my personal friendship and affection for him, but because I think he well deserves such recognition.

And had I thought right to do so, I felt sure that you would have fully appreciated my motives, and that it would have done no injury to our friendship.

But as I told the Council I did not think this a case where either of you had any right to be excluded by the other. I told them that had Forbes been first named, I should have thought it injudicious to bring you forward, and that, as you were named, I for my own part should not have brought forward Forbes as a candidate ; that therefore while willing to speak up to any extent for Forbes' *positive* merits and deserts, I would carefully be understood to give no opinion as to your and his *relative* standing.

They did not take much by my speech therefore either way, more especially as I voted for *both* of you.

I hate doing anything of the kind "unbeknownst" to people, so there is the exact history of my proceedings. If I had been able to come to the clear conclusion that the claims of either of you were strongly superior to those of the other, I think I should have had the honesty and moral courage to "act accordin'," but I really had not, and so there was no part to play but that of a sort of Vicar of Bray.—Ever yours faithfully,

T. H. Huxley.

Forbes' reply was a letter which Huxley, after his friend's death, held "among his most precious possessions." It appeared without names in the obituary notice of Forbes in the *Literary Gazette* for November 25, 1854, as an example of his unselfish generosity :—

I heartily concur in the course you have taken, and had I been placed as you have been, would have done exactly the same. . . . Your way of proceeding was as true an act of friendship as any that could be performed. As to myself, I dream so little about medals, that the notion of being on the list never entered my brain, even when asleep. If it ever comes I shall be pleased and thankful ; if it does not, it is not the sort of thing to break my equanimity. Indeed, I would always like to see it given not as a mere honour, but as a help to a good man, and this it is assuredly in Hooker's case. Government people are so ignorant that they require to have merits drummed into their heads by all possible means, and Hooker's getting the medal may be of real service to him before long. I am in a snug, though not an idle, nest,—he has not got his resting-place yet. And so, my dear Huxley, I trust that you know me too well to think that I am either grieved or envious, and you, Hooker, and I are much of the same way of thinking.

It is interesting to record the same scrupulosity over the election to the Registrarship of the University of London in 1856, when, having begun to canvass for Dr. Latham before his friend Dr. W. B. Carpenter entered the field, he writes to Hooker :—

I at once, of course, told Carpenter precisely what I had done. Had I known of his candidature earlier, I should certainly have taken no active part on either side —not for Latham, because I would not oppose Carpenter, and not for Carpenter, because his getting the Registrarship would probably be an advantage for me, as I should have a good chance of obtaining the Examinership in Physiology and Comparative Anatomy which he would vacate. Indeed, I refused to act for Carpenter in a case in which he asked me to do so, partly for this reason and partly because I felt thoroughly committed to Latham. Under these circumstances I think you are quite absolved from any pledge to me. It's deuced hard to keep straight in this wicked world, but as you say the only chance is to out with it, and I thank you much for writing so frankly about the matter. I hope it will be as fine as to-day at Down.[1]

_____

[1] Charles Darwin's home in Kent.

Unfortunately the method was not so successful with smaller minds. Once in 1852, when he had to report unfavourably on a paper for the *Annals of Natural History* on the structure of the Starfishes, sent in by an acquaintance, he felt it right not to conceal his action, as he might have done, behind the referee's usual screen of anonymity, but to write a frank account of the reasons which had led him so to report, that he might both clear himself of the suspicion of having dealt an unfair blow in the dark, and give his acquaintance the opportunity of correcting and enlarging his paper with a view of submitting it again for publication.

In this case the only result was an impassioned correspondence, the author even going so far as to suggest that Huxley had condemned the paper without having so much as dissected an Echinoderm in his life! and then all intercourse ceased, till years afterwards the gentleman in question realised the weaknesses of his paper and repented him of his wrath.

Before leaving London to begin his work at Tenby as Naturalist to the Survey, he delivered at St. Martin's Hall, on July 22, an address on the "Educational Value of the Natural History Sciences."[1] This, when it came out later as a

---

[1] The subsequent reference is to the words, "I cannot but think that he who finds a certain proportion of pain and evil inseparably woven up in the life of the very worms will bear his own share with more courage and submission ; and will, at any rate, view with suspicion those weakly amiable theories of the divine govern-

pamphlet, he sent to his Tenby friend Dr. Dyster (of whom hereafter), to whose criticism on one passage he replied on October 10 :—

> . . . I am rejoiced you liked my speechment. It was written hastily and is, like its speaker, I fear, more forcible than eloquent, but it can lay claim to the merit of being sincere.
>
> My intention on p. 28 was by no means to express any satisfaction at the worms being as badly off as ourselves, but to show that pain being everywhere is inevitable, and therefore like all other inevitable things to be borne. The rest of it is the product of my scientific Calvinism, which fell like a shell at your feet when we were talking over the fire.
>
> I doubt, or at least I have no confidence in, the doctrine of ultimate happiness, and I am more inclined to look the opposite possibility fully in the face, and if that also be inevitable, make up my mind to bear it also.
>
> You will tell me there are better consolations than Stoicism ; that may be, but I do not possess them, and I have found my "grin and bear it" philosophy stand me in such good stead in my course through oceans of disgust and chagrin, that I should be loth to give it up.

The summer of 1854 was spent in company with the Busks at Tenby, amid plenty of open-air work and in great peace of mind, varied with a short visit to Liverpool in order to talk business with his friend

ment, which would have us believe pain to be an oversight and a mistake, to be corrected by and by." (*Collected Essays*, iii. p. 62.) This essay contains the definition of science as "trained and organised common sense," and the reference to a new "Peter Bell" which suggested Miss May Kendall's spirited parody of Wordsworth :—

> Primroses by the river's brim
> Dicotyledons were to him,
> And they were nothing more.

Forbes, who was eager that Huxley should join him in Edinburgh.

TENBY, SOUTH WALES, *Sept.* 3, 1854.

I have been here since the middle of August getting rid of my yellow face and putting on a brown one, banishing dyspepsias and hypochondrias and all such other town afflictions to the four winds, and rejoicing exceedingly that I am out of the way of that pest, the cholera, which is raging just at present in London.

After I had arranged to come here to do a lot of work of my own which can only be done by the seaside, our Director, Sir Henry de la Beche,[1] gave me a special mission of his own whereby I have the comfort of having my expenses paid, but at the same time get it it taken out of me in additional labour, so my recreation is anything but leisure.

Oct. 14.

I left this place for a week's trip to Liverpool in the end of September. The meeting of the British Association was held there, but I went not so much to be present as to meet Forbes, with whom I wanted to talk over many matters concerning us both. Forbes had a proposition that I should go to Edinburgh to take part of the duties of the Professor of Physiology there, who is in bad health, with the ultimate aim of succeeding to the chair. It was a tempting offer made in a flattering manner, and presenting a prospect of considerably better emolument than my special post, but it had the disadvantage of being but an uncertain position. Had I accepted, I should have been at the mercy of the actual Professor—and that is a position I don't like standing in, even with the best of men, and had he died or resigned at any time the Scotch chairs are

---

[1] 1796-1855. His idea of a Geological Survey based on the Ordnance Survey was taken up by Government in 1832. To his initiative also was due the Museum of Practical Geology in Jermyn Street, and later the School of Mines.

so disposed of that there would have been nothing like a certainty of my getting the post, so I definitely declined —I hope wisely.

After some talk, Forbes agreed with my view of the case, so he is off to Edinburgh, and I shall go off to London. I hope to remain there for my life long.

He had long felt that London gave the best opportunities for a scientific career, and it was on his advice that Tyndall had left Queenwood College for the Royal Institution, where he was elected Professor of Natural Philosophy in 1853 :—

6 UPPER YORK PLACE, ST. JOHN'S WOOD,
Feb. 25, 1853.

MY DEAR TYNDALL—Having rushed into more responsibility than I wotted of, I have been ruminating and taking counsel what advice to give you. When I wrote I hardly knew what kind of work you had in your present office, but Francis has since enlightened me. I thought you had more leisure. One thing is very clear —you must come out of that. Your Pegasus is quite out of place, ploughing. You are using yourself up in work that comes to nothing, and so far as I can see cannot be worse off.

Now what are your prospects? Why, as I told you before, you have made a *succès* here and must profit by it. The other night your name was mentioned at the Philosophical Club (the most influential scientific body in London) with great praise. Gassiot, who has great influence, said in so many words, "you had made your fortune," and I frankly tell you I believe so too, if you can only get over the next three years. So you see that *quoad* position, like Quintus Curtius, there is a "fine opening" ready for you, only mind you don't spoil it by any of your horrid modesty.

So much for glory—now for economics. I have been

trying to ferret out more nearly your chances of a post, and here are my results (which, I need not tell you, must be kept to yourself).

At the Museum in Jermyn Street, Playfair, Forbes, Percy,[1] and I think Sir Henry would do anything to get you, and eliminate —— ; but, so far as I can judge, the probability of his going is so small that it is not worth your while to reckon upon it. Nevertheless it may be comforting to you to know that in case of anything happening these men will help you tooth and nail. Cultivate Playfair when you have a chance—he is a good fellow, wishes you well, has great influence, and will have more. *Entre nous,* he has just got a new and important post under Government.

Next, the Royal Institution. This is where, as I told you, you ought to be—looking to Faraday's place. Have no scruple about your chemical knowledge ; you won't be required to train a college of students in abstruse analyses ; and if you were, a year's work would be quite enough to put you at ease. What they want, and what you have, are *clear powers of exposition*—so clear that people may think they understand even if they don't. That is the secret of Faraday's success, for not a tithe of the people who go to hear him really understand him.

However, I am afraid that a delay must occur before you can get placed at the Royal Institution, as you cannot hold the Professorship until you have given a course of lectures there, and it would seem that there is no room for you this year. However, I must try and learn more about this.

Under these circumstances the London Institution looks tempting. I have been talking over the matter with Forbes, whose advice I look upon as first-rate in all these things, and he is decidedly of the opinion that you

---

[1] Dr. John Percy, 1817-1889, Professor of Metallurgy at the Royal School of Mines from 1851 till its removal to South Kensington in 1879.

should take the London Institution if it is offered you.
He says that lecturing there and lecturing at other
Institutions, and writing, you could with certainty make
more than you at present receive, and that you would
have the command of a capital laboratory and plenty of time.

Then as to position—of which I was doubtful—it
appears that Grove [1] has made it a good one.

It is of great importance to look to this point in
London—to be unshackled by anything that may prevent
you taking the highest places, and it was only my fear
on this head that made me advise you to hesitate about
the London Institution.   More consideration leads me to
say, take that, if it will bring you up to London at once,
so that you may hammer your reputation while it is hot.

However, consider all these things well, and don't be
hasty.   I will keep eyes and ears open and inform you
accordingly.   Write to me if there is anything you want
done, supposing always there is nobody who will do it
better—which is improbable.—Ever yours,

T. H. HUXLEY.

But this year of victory was not to pass away
without one last blow from fate.   On November 18,
Edward Forbes, the man in whom Huxley had found
a true friend and helper, inspired by the same ideals
of truth and sincerity as himself, died suddenly at
Edinburgh.   The strong but delicate ties that united
them were based not merely upon intellectual affinity,
but upon the deeper moral kinship of two strong
characters, where each subordinated interest to ideal,
and treated others by the measure of his own self-
respect.   As early as March 1851 he had written :—

[1] Sir William Robert Grove, 1811-1896, the physicist and
lawyer, a judge 1871-87, who established the correlation of
physical forces, 1846.

I wish you knew my friend Prof. Forbes. He is the best creature you can imagine, and helps me in all manner of ways. A man of very great knowledge, he is wholly free from pedantry and jealousy, the two besetting sins of literary and scientific men. Up to his eyes in work, he never grudges his time if it is to help a friend. He is one of the few men I have ever met to whom I can feel obliged, without losing a particle of independence or self-respect.

The following from a letter to Hooker, announcing Forbes' death, is a striking testimony to his worth :—

I think I have never felt so crushed by anything before. It is one of those losses which cannot be replaced either to the private friend or to science. To me especially it is a bitter loss. Without the aid and sympathy he has always given me from first to last, I should never have had the courage to persevere in the course I have followed. And it was one of my greatest hopes that we should work in harmony for long years at the aims so dear to us both.

But it is otherwise, and we who remain have nothing left but to bear the inevitable as we best may.

And again a few days later :—

I have had no time to write to you again till now, but I write to say how perfectly you express my own feeling about our poor friend. One of the first things I thought of was that medal business,[1] and I never rejoiced in anything more than that I had not been deterred by any moral cowardice from acting as I did.

As it is I reckon that letter (which I will show you some day) among my most precious possessions.

Huxley's last tribute to his dead friend was the organising a memorial fund, part of which went to

[1] P. 160 *sq.*

getting a bust of him made, part to establishing an
Edward Forbes medal, to be competed for by the
students of his old school in Jermyn Street.

As Huxley had been Forbes' successor at Jermyn
Street, so now he seemed to many marked out to
succeed him at Edinburgh. In November he writes
to Hooker :—

People have been at me about the Edinburgh chair.
If I could contrive to stop here, between you and I, I
would prefer it to half a dozen Edinburgh chairs, but
there is a mortal difference between £200 and £1000
a year. I have written to say that if the Professors can
make up their minds they wish me to stand, I will—
if not, I will not. For my own part, I believe my
chances would be very small, and I think there is every
probability of their dividing the chair, in which case
I certainly would not go. However, I hate thinking
about the thing.

And also to his sister :—

Nov. 26, 1854.

MY DEAREST LIZZIE—I feel I have been silent very
long—a great deal too long—but you would understand
if you knew how much I have to do ; why, with every
disposition to do otherwise, I now write hardly any but
business letters. Even Nettie comes off badly, I am
afraid. When a man embarks as I have done, with
nothing but his brains to back him, on the great sea of
life in London, with the determination to *make* the
influence and the position and the money which he hasn't
got, you may depend upon it that the fierce wants and
interests of his present and immediate circle leave him
little time to think of anything else, whatever old loves
and old memories may be smouldering as warmly as ever
below the surface. So, sister mine, you must not imagine

because I do not write that therefore I do not think of
you or care to know about you, but only that I am eaten
up with the zeal of my own house, and doing with all
my heart the thing that the moment calls for.

The last year has been eventful for me. There is
always a Cape Horn in one's life that one either weathers
or wrecks one's self on. Thank God, I think I may say I
have weathered mine—not without a good deal of damage
to spars and rigging though, for it blew deuced hard on
the other side.

At the commencement of this year my affairs came to
a crisis. The Government, notwithstanding all the representations which were made to them, would neither give
nor refuse the grant for the publication of my work, and
by way of cutting short all further discussion the
Admiralty called upon me to serve. A correspondence
ensued, in which, as commonly happens in these cases,
they got the worst of it in logic and words, and I in
reality and "tin." They answered my syllogism by the
irrelevant and absurd threat of stopping my pay if I did
not serve at once. Here was a pretty business ! However, it was no use turning back when so much had been
sacrificed for one's end, so I put their Lordships' letter up
on my mantelpiece and betook myself to scribbling for
my bread. They, on the other hand, removed my name
from the List. So there was an interregnum when I was
no longer in Her Majesty's service. I had already joined
the *Westminster Review,* and had inured myself to the
labour of translation—and I could get any amount of
scientific work I wanted—so there was a living, though
a scanty one, and amazingly hard work for it. My pen
is not a very facile one, and what I write costs me a good
deal of trouble.

In the spring of this year, however, a door opened.
My poor lost friend Professor Forbes—whose steady
attachment and aid had always been of the utmost service
to me—was called to fill the chair of Natural History in

Edinburgh at a moment's notice. It is a very valuable appointment, and he was obliged to fill it at once. Of course he left a number of vacancies behind, among them one at the Government School of Mines in Jermyn Street, where he lectured on Natural History. I was called upon to take up his lectures where he left off, in the same sudden way, and the upshot of it all was that I became permanently attached—with £200 a year pay. In other ways I can make a couple of hundred a year more even now, and I hope by and by to do better. In fact, a married man, as I hope soon to be, cannot live at all in the position which I ought to occupy under less than six hundred a year. If I keep my health, however, I have every hope of being able to do this—but, as the jockeys say, the pace is severe. Nettie is coming over in the spring, and if I have any luck at all, I mean to have paid off my debts and to be married by this time next year.[1]

In the meanwhile, strangely enough—and very painfully for me—new possibilities have sprung up. My poor friend Forbes died only a week ago, just as he was beginning his course and entering upon as brilliant a career as ever was opened to any scientific man in this country.

I cannot tell you how deeply this has shocked me. I owe him so much, I loved him so well, and I have so

---

[1] He writes on July 21, 1851:—"I commenced life upon nothing at all, and I had to borrow in the ordinary way from an agent for the necessary expenses of my outfit. I sent home a great deal of money, but notwithstanding, from the beautiful way they have of accumulating interest and charges of one description and another, I found myself £100 in debt when I returned—besides something to my brother, about which, however, I do not suppose I need trouble myself just at present. As you may imagine, living in London, my pay now hardly keeps me, to say nothing of paying off my old scores. I could get no account of how things were going on with my agent while I was away, and therefore I never could tell exactly how I stood."

very very few friends in the true sense of the word, that it has been perhaps a greater loss to me than to any one —although there never was a man so widely lamented. One could trust him so thoroughly! However, he has gone, poor fellow, and there is nothing for it but to shut one's self up again—and I was only going to say that his death leaves his post vacant, and I have been strongly urged to become a candidate for it by several of the most influential Edinburgh Professors. I am greatly puzzled what to do. I do not want to leave London, nor do I think much of my own chances of success if I become a candidate—though others do. On the other hand, a stipend which varies between £800 and £1200 a year is not to be pooh-poohed.

We shall see. If I can carry out some arrangements which are pending with the Government to increase my pay to £400 a year, I shall be strongly tempted to stop in London. It is *the* place, the centre of the world.

In the meanwhile, as things always do come in heaps, I obtained my long-fought-for Grant—though indirectly —from the Government, which is, I think, a great triumph and vindication of the family motto—*tenax propositi*. Like many long-sought-for blessings, however, it is rather a bore now I have it, as I don't see how I am to find time to write the book. But things "do themselves" in a wonderful way. I'll tell you how many irons I have in the fire at this present moment :—(1) a manual of Comparative Anatomy for Churchill; (2) my "Grant" book; (3) a book for the British Museum people (half done); (4) an article for Todd's *Cyclopædia* (half done); (5) sundry memoirs on Science; (6) a regular Quarterly article in the *Westminster*; (7) lectures at Jermyn Street in the School of Mines; (8) lectures at the School of Art, Marlborough House; (9) lectures at the London Institution, and odds and ends. Now, my dearest Lizzie, whenever you feel inclined to think it unkind I don't write, just look at that list, and remember that all these things require strenuous attention and concentration of the faculties, and leave one not very fit for anything else. You will say that it is bad to be so entirely absorbed in these things, and to that I heartily say Amen !—but you might as well argue with a man who has just mounted the favourite for the "Oaks" that it is a bad thing to ride fast. He admits that, and is off like a shot when the bell rings nevertheless. My bell has rung some time, and thank God the winning post is in sight.

Give my kindest regards to the doctor and special love to all the children. I send a trifle for my godson and some odds and ends in the book line, among other things a Shakespeare for yourself, dear Liz.—Believe me, ever your affec. brother,   T. H. HUXLEY.

In December the Edinburgh chair was practically offered to him undivided; but by that time the London authorities thought they had better make it worth his while to stay at Jermyn Street, and with negotiations begun for this end he refused to stand for Edinburgh. In the following spring, however, he was again approached from Edinburgh—not so much to withdraw his refusal and again become a candidate, as to let it be made known that he would accept the chair if it were offered him. But his position in London was now established; and he preferred to live in London on a bare sufficiency rather than to enjoy a larger income away from the centre of things.

Two letters to Tyndall, which refer to the division of labour in the science reviews for the *Westminster* (see p. 124), indicate very clearly the high pressure at which Huxley had already begun to work :—

TENBY, SOUTH WALES,
Oct. 22, 1854.

MY DEAR TYNDALL—I was rejoiced to find you entertaining my proposition at all. No one believes how hard you work more than I, but I was not going to be such a bad diplomatist as to put that at the head of my letter, and if I had thought that what I want you to do involved any great accession thereto, I think I could not have mustered up the face to ask you. But really and truly, so long as it is confined to our own department it is no great affair. You make me laugh at the long face you pull about the duties, based on my phrase. The fact is, you notice what you like, and what you do not you leave undone, unless you get an editorial request to say something about a particular book. The whole affair is entirely in your own hands—at least it is in mine—as I went upon my principle of having a row at starting. . . .

Now here is an equitable proposition. Look at my work. I have a couple of monographs, odds and ends of papers for journals, a manual and some three courses of lectures to provide for this winter. "My necessities are as great as thine," as Sir Philip Sidney didn't say, so be a brick, split the difference, and say you will be ready for the April number. I will write and announce the fact to Chapman.

What idiots we all are to toil and slave at this pace. I almost repent me of tempting you—after all—so I promise to hold on if you really think you will . be overdoing it.

With you I envy Francis his gastric energies. I feel I have done for myself in that line, and am in for a lifelong dyspeps. I have not, now, nervous energy enough for stomach and brain both, and if I work the latter, not even the fresh breezes of this place will keep the former in order. That is a discovery I have made here, and though highly instructive, it is not so pleasant as some other physiological results that have turned up.

Chapman, who died of cholera, was a distant relative of my man. The poor fellow vanished in the middle of an unfinished article, which has appeared in the last *Westminster*, as his forlorn *vale !* to the world. After all, that is the way to die, better a thousand times than drivelling off into eternity betwixt awake and asleep in a fatuous old age.—Believe me, ever yours faithfully,

T. H. HUXLEY.

On Tyndall consenting, he wrote again on the 29th :—

I rejoice in having got you to put your head under my yoke, and feel ready to break into a hand gallop on the strength of it.

I have written to Chapman to tell him you only make an experiment on your cerebral substance, whose continuance depends on tenacity thereof.

I didn't suspect you of being seduced by the magnificence of the emolument, you Cincinnatus of the laboratory. I only suggested that as pay sweetens labour, *a fortiori* it will sweeten what to you will be no labour.

I'm not a miserable mortal now—quite the contrary. I never am when I have too much to do, and my sage reflection was not provoked by envy of the more idle. Only I do wish I could sometimes ascertain the exact *juste milieu* of work which will suit, not my head or will, *these* can't have too much ; but my absurd stomach.

The Edinburgh candidature, the adoption of his wider scheme for the carrying out of the coast survey, and his approaching marriage, are touched upon in the following letters to Dr. Frederick Dyster [1] of Tenby, whose keen interest in marine zoology was the starting-point of a warm friendship with the rising naturalist, some fifteen years his junior. He was strongly urged by the younger man to complete and systematise his observations by taking in turn all the species of each genus of annelids found at Tenby, and working them up into a series of little monographs "which would be the best of all possible foundations for a History of the British Annelidæ."

To Dr. Dyster

Jan. 5, 1855.

[He begins by confessing "a considerable liberty" he had been taking with Dyster's name, in calling a joint discovery of theirs, which he described in the *Edinburgh New Philosophical Journal*, Protula Dysteri.]

Are you very savage? If so, you must go and take a walk along the sands and see the slant rays of the sunset tipping the rollers as they break on the beach; that always made even *me* at peace with all the world, and *a fortiori* it will you.

Truly, I wish I had any such source of consolation. Chimney pots are highly injurious to my morals, and my temper is usually in proportion to the extent of my horizon.

---

[1] It was to Dyster that Huxley owed his introduction in 1854 to F. D. Maurice (whose work in educating the people he did his best to help), and later to Charles Kingsley, whom he first met at the end of June 1855, "What Kingsley do you refer to?" he writes on May 6, "*Alton Locke* Kingsley or Photographic Kingsley? I shall be right glad to find good men and true anywhere, and I will take your bail for any man. But the work must be critically done."

I have been swallowing oceans of disgust lately. All sorts of squabbles, some made by my own folly and others by the malice of other people, and no great sea and sky to go out under, and be alone and forget it all.

You may have seen my name advertised by Reeve[1] as about to write a memoir of poor Forbes, to be prefixed to a collection of his essays. I found that to be a mere bookseller's dodge on Reeve's part, and when I made the discovery, of course we had a battle-royal, and I have now wholly withdrawn from it.

I find, however, that one's kind and generous friends imagine it was an electioneering manœuvre on my part for Edinburgh. Imagine how satisfactory. I forget whether I told you that I had been asked to stand for Edinburgh and have done so. Whether I shall be appointed or not I do not know. So far as my own wishes go, I am in a curiously balanced state of mind about it. Many things make it a desirable post, but I dread leaving London and its freedom—its Bedouin sort of life—for Edinburgh and no whistling on Sundays. Besides, if I go there, I shall have to give up all my coast-survey plans, and all their pleasant concomitants.

Apropos of Edinburgh I feel much like the Irish hodman who betted his fellow he could not carry him up to the top of a house in his hod. The man did it, but Pat turning round as he was set down on the roof, said, " Ye've done it, sure enough, but, bedad, I'd great hopes ye'd let me fall about three rounds from the top." Bedad, I'm nearly at the top of the Scotch ladder, but I've hopes.

It is finally settled that the chair will not be divided. I told them frankly I would not go if it were.

Has Highly sent your books yet?—Ever yours faithfully,   T. H. HUXLEY.

[1] Henry Reeve, 1813-1895, editor of the *Edinburgh Review* from 1855 to his death. He translated De Tocqueville, and wrote or edited several works on art, letters, and biography.

JERMYN STREET, *Feb.* 13, 1855.

MY DEAR DYSTER—. . . I will do my best to help ——to some alumni if the chance comes in my way, though, as you say, I don't like him. I can't help it. I respect piety, and hope I have some after my own fashion, but I have a profound prejudice against the efflorescent form of it. I never yet found in people thoroughly imbued with that pietism, the same notions of honour and straight-forwardness that obtain among men of the world. It may be otherwise with ——, but I can't help my pagan prejudice. So don't judge harshly of me thereanent.

About Edinburgh, I have been going to write to you for days past. I have decided on withdrawing from the candidature, and have done so. In fact the more I thought of it the less I liked it. They require nine months' lectures some four or five times a week, which would have thoroughly used me up, and completely put a stop to anything like original work ; and then there was a horrid museum to be arranged, work I don't care about, and which would have involved an amount of intriguing and heart-burning, and would have required an amount of diplomacy to carry to a successful issue, for which my temper and disposition are wholly unfitted.

And then I felt above all things that it was for me an imposture. Here have I been fighting and struggling for years, sacrificing everything to be a man of science, a genuine worker, and if I had obtained the Edinburgh chair, I should have been in reality a mere pedagogue and a man of science only in name. Such were my notions, and if I hesitated at all and allowed myself to become a candidate, it was only because I have other interests to consult than my own. Intending to " range myself " one of these days and become a respectable member of society, I was bound to consider my material

interests. And so I should have been still a candidate
for Edinburgh had not the Government here professed
themselves unwilling to lose my services, adding the
"material guarantee" of an addition to my income,
which, though by no means bringing it up to the point
of Edinburgh, will still enable me (*das heisst* "us") to
live comfortably here.

I must renounce the "pomps and vanities," but all
those other "lusts of the flesh" which may beseem a
gentleman may be reasonably gratified.

Don't you think I have been wise in my Hercules
choice ? After all I don't lay claim to any great merit,
seeing it was anything but certain I should get Edinburgh.

The best of all is that I have every reason to believe
that Government will carry out my scheme for a coast
survey, so happily and pleasantly begun at Tenby last
year.

The final arrangements are almost complete, and I
believe you may make up your mind to have four
months of me next year. Tenby shall be immortalised
and Jenkyn [1] converted into a philosopher. By the way,
I think the best way would be to retain the shells till I
come. My main purpose is to have in them a catalogue
of what Tenby affords.

Pray give my kind remembrances to Mrs. Dyster,
and believe me, ever yours,   T. H. HUXLEY.

April 1, 1855.

MY DEAR DYSTER—By all that's good, your last note,
which lies before me, has date a month ago. I looked at
it just now, and became an April fool on the instant.

All the winds of March, however, took their course
through my thorax and eventuated in lectures. At least
that is all the account I can give to myself of the time,

---

[1] Jenkyn was employed to collect shells, etc., at Tenby. He
is often alluded to as "the Professor."

and an unprofitable account it is, for everything but
one's exchequer.

So far as knowledge goes it is mere prodigality
spending one's capital and adding nothing, for I find
the physical exertion of lecturing quite unfits me for
much else. Fancy how last Friday was spent. I went
to Jermyn Street in the morning with the intention of
preparing for my afternoon's lecture. People came
talking to me up to within a quarter of an hour of the
time, so I had to make a dash without preparation.
Then I had to go home to prepare for a second lecture in
the evening, and after that I went to a soirée, and got
home about one o'clock in the morning.

I go on telling myself this won't do, but to no purpose.

You will be glad to hear that my affairs here are
finally settled, and I am regularly appointed an officer
of the survey with the commission to work out the
natural history of the coast.

Edinburgh has been tempting me again, and in fact
I believe I was within an ace of going there, but the
Government definitely offering me this position, I was
too glad to stop where I am.

I can make six hundred a year here, and that being
the case, I conceive I have a right to consult my own
inclinations and the interests of my scientific reputation.
The coast survey puts in my hands the finest opportunities
that ever a man had, and it is a pity if I do not make
myself something better than a Caledonian pedagogue.

The great first scheme I have in connection with my
new post is to work out the Marine Natural History of
Britain, and to have every species of sea beast properly
figured and described in the reports which I mean from
time to time to issue. I can get all the engravings and
all the printing I want done, but of course I am not so
absurd as to suppose I can work out all these things
myself. Therefore my notion is to seek in all highways
and byways for fellow-labourers. Busk will, I hope,

supply me with figures and descriptions of the British Polyzoa and Hydrozoa, and I have confidence in my friend, Mr. Dyster of Tenby (are you presumptuous enough to say you know him ?) for the Annelids, if he won't object to that mode of publishing his work. The Mollusks, the Crustaceans, and the Fishes, the Echinoderms and the Worms, will give plenty of occupation to the other people, myself included, to say nothing of distribution and of the recent geological changes, all of which come within my programme.

Did I not tell you it was a fine field, and could the land o' cakes give me any scope like this ?

April 9, 1855.

MY DEAR DYSTER—I didn't by any means mean to be so sphinx-like in my letter, though you have turned out an Œdipus of the first water. True it is that I mean to " range myself," " live cleanly and leave off sack," within the next few months—that is to say, if nothing happen to the good ship which is at present bearing my fiancée homewards.

So far as a restless mortal—more or less aweary of most things—like myself can be made happy by any other human being, I believe your good wishes are safe of realisation ; at any rate, it will be my fault if they are not, and I beg you never to imagine that I could confound the piety of friendship with the " efflorescent " variety.

I hope to marry in July, and make my way down to Tenby shortly afterwards, and I am ready to lay you a wager that your vaticinations touching the amount of work that *won't* be done don't come true.

So much for wives—now for *worms*—(I could not for the life of me help the alliteration). I, as right reverend father in worms and Bishop of Annelidæ, do not think I ought to interfere with my most promising son, when a channel opens itself for the publication of his labours. So do what you will *apropos* of J——. If he does not do the worms any better than he did the zoophytes, he won't interfere with my plans.

I shall be glad to see Mrs. Buckland's Echinoderm. I think it must be a novelty by what you say. She is a very jolly person, but I have an unutterable fear of scientific women.—Ever yours,   T. H. HUXLEY.

May 6, 1855.

My ship is not come home but is coming, and I have been in a state of desperation at the continuous east winds. However, to-day there is a westerly gale, and if it lasts I shall have news soon. You may imagine that I am in an unsatisfactory state of mind between this and lecturing five times a week.

I beg to say that the "goods" I expect are home produce transplanted (or sent a voyage as you do Madeira), and not foreign growth by any means. But it is five years since we met, I am another man altogether, and if my wife be as much altered, we shall need a new introduction. Correspondence, however active, is a poor substitute for personal communication and tells one but little of the inner life.

Finally, on the eve of his marriage in July, Tyndall congratulates him on being appointed to deliver the next course of Fullerian Lectures at the Royal Institution :—

The fates once seemed to point to our connection in a distant land : we are now colleagues at home, and I can claim you as my scientific brother. May the gods continue to drop fatness upon you, and may your next great step be productive of all the felicity which your warmest friends or your own rebellious heart can desire.

CHAPTER IX

1855

MISS HEATHORN and her parents reached England at
the beginning of May 1855, and took up their abode
at 8 Titchfield Terrace, not far from Huxley's own
lodgings and his brother's house. One thing, however, filled Huxley with dismay. Miss Heathorn's
health had broken down utterly, and she looked at
death's door. All through the preceding year she
had been very ill; she had gone with friends, Mr.
and Mrs. Wise, to the newly-opened mining-camp at
Bathurst, and she and Mrs. Wise were indeed the
first women to visit it; returning to Sydney after
rather a rough time, she caught a chill, and being
wrongly treated by a doctor of the blood-letting,
calomel-dosing school, she was reduced to a shadow,
and only saved by another practitioner, who reversed
the treatment just in time.

In his letters to her, Huxley had not at first
realised the danger she had been in; and afterwards
tried to keep her spirits up by a cheerful optimism
that would only look forward to their joyful union

and many years of unbroken happiness to atone for
their long parting.

But the reality alarmed him.   He took her to one
of the most famous doctors of the day, as if merely a
patient he was interested in.   Then as one member of
the profession to another, he asked him privately his
opinion of the case.   "I give her six months of life,"
said Æsculapius.   "Well, six months or not," replied
Huxley, "she is going to be my wife."   The doctor
was mightily put out.   "You ought to have told me
that before."   Of course, the evasive answer in such
a contingency was precisely what Huxley wished to
avoid.   Happily another leading doctor held a much
more favourable opinion, and said that with care her
strength would come back, slowly but surely.

14 WAVERLEY PLACE, *Wednesday.*

MY DEAR HOOKER—My wife and I met again on
Sunday last, and I have established herself, her father and
mother, close by me here at 8 Titchfield Terrace, Regent's
Park, and whenever you and Mrs. Hooker are in this part
of the world, and can find time to call there, you will find
her anything but surprised to see you.

God help me ! I discover that I am as bad as any
young fool who knows no better, and if the necessity for
giving six lectures a week did not sternly interfere, I
should be hanging about her ladyship's apron-strings all
day.   She is in very bad health, poor child, and I have
some reason to be anxious, but I have every hope she will
mend with care.

Oh this life ! "atra cura," as old Thackeray has it,
sits on all our backs and mingles with all our happiness.

But if I go on talking in this way you will wonder what
has come over my philosophership.—Ever yours,
                                   T. H. HUXLEY.

Black Care was still in the background, but had
relaxed her hold upon him. His spirits rose to the
old point of gaiety. He writes how he gives a lively
lecture to his students, and in the midst of it Satan
prompts him to crow or howl—a temptation happily
resisted. He makes atrocious puns in bidding Hooker
to the wedding, which took place on July 21.

                      JERMYN STREET, *July 6*, 1855.

MY DEAR HOOKER — I ought long since to have
thanked you in Thomson's name as well as my own for
your *Flora Indica.* Some day I promise myself much
pleasure and profit from the digestion of the Introductory
Essay, which is probably as much as my gizzard is competent to convert into nutrition.

I terminate my Baccalaureate and take my degree of
M.A.trimony (isn't that atrocious ?) on Saturday, July 21.
After the unhappy criminals have been turned off, there
will be refreshment provided for the sheriffs, chaplain,
and spectators. Will you come ? Don't if it is a bore,
but I should much like to have you there.

It was not a large party that assembled at the
George Huxleys for the wedding, but all were lifelong friends, including, besides the Fanning clan and
Mrs. Griffiths, an old Australian ally, Hooker,
Tyndall, and Dr. and Mrs. Carpenter. There was
none present but felt that abundant happiness was at
least well earned after eight years of trial, and still
more that its best guarantee was the firm loyalty

and devotion that had passed through so many dangers of absence and isolation, so many temptations to renounce the ideal course under stress of circumstance, only to emerge strengthened and ennobled by the stern discipline of much sacrifice.

Great as was his new happiness, he hardly stood in need of Darwin's word of warning: "I hope your marriage will not make you idle; happiness, I fear, is not good for work." Huxley could not sit idle for long. If he had no occupation on hand, something worth investigation—and thorough investigation—was sure to catch his eye. So he writes to Hooker from Tenby :—

15 ST. JULIAN'S TERRACE, TENBY,
Aug. 16, 1855.

MY DEAR HOOKER—I am so near the end of the honeymoon that I think it can hardly be immodest if I emerge from private life and write you a letter, more particularly as I want to know something. I went yesterday on an expedition to see the remains of a forest which exists between tidemarks at a place called Amroth, near here.

So far as I can judge, there can be no doubt that this really is a case of downward movement. The stools of the trees are in their normal position, and their roots are embedded and interwoven in a layer of stiff blue clay, which lies immediately beneath the superficial mud of the shore. Layers of leaves, too, are mixed up with the clay in other parts, and the bark of some of the trees is in perfect preservation. The condition of the wood is very curious. It is like very hard cheese, so that you can readily cut slices with a spade, and yet where more of the trunk has been preserved some parts are very hard. The trees are, I fancy, Beech and Oak. Could you identify slices if I were to send you some ?

Now it seems to me that here is an opportunity one does not often have of getting some information about the action of sea water on wood, and on the mode in which these vegetable remains may become embedded, etc., etc., and I want to get you to tell me where I can find information on submerged forests in general, so as to see to what points one can best direct one's attention, and to suggest any inquiries that may strike yourself.

I do not see how the stumps can occur in this position without direct sinking of the land, and that such a sinking should have occurred tallies very well with some other facts which I have observed as to the nature of the bottom at considerable depths here.

We had the jolliest cruise in the world by Oxford, Warwick, Kenilworth, Stratford, Malvern, Ross, and the Wye, though it *was* a little rainy, and though my wife's strength sadly failed at times.

Still she was on the whole much better and stronger than I had any right to expect, and although I get frightened every now and then, yet there can be no doubt that she is steadily though slowly improving. I have no fears for the ultimate result, but her amendment will be a work of time. We have really quite settled down into Darby and Joan, and I begin to regard matrimony as the normal state of man. It's wonderful how light the house looks when I come back weary with a day's boating to what it used to do.

I hope Mrs. Hooker is well and about again. Pray give her our very kind regards, and believe me, my dear Hooker, ever yours,   T. H. HUXLEY.

At Tenby he stayed on through August and September, continuing his occupations of the previous summer, dredging up specimens for his microscope, and working partly for his own investigations, partly for the Geological Survey.

CHAPTER X

1855–1858

Up to his appointment at the School of Mines, Huxley's work had been almost entirely morphological, dealing with the Invertebrates. His first investigations, moreover, had been directed not to species-hunting, but to working out the real affinities of little known orders, and thereby evolving a philosophical classification from the limbo of "Vermes" and "Radiata."

He had continued the same work by tracing homologies of development in other classes of animals, such as the Cephalous Mollusca, the Articulata, and the Brachiopods. On these subjects, also, he had a good deal of correspondence with other investigators of the same cast of mind, and even when he did not carry conviction, the impression made by his arguments may be judged from the words of Dr. Allman, no mean authority, in a letter of May 2, 1852 :—

I have thought over your arguments again and again, and while I am the more convinced of their ingenuity, originality, and *strength*, I yet feel ashamed to confess that I too must exclaim "tenax propositi." When was it otherwise in controversy?

Other speculations arising out of these researches had been given to the public in the form of lectures, notably that on Animal Individuality at the Royal Institution in 1852.

But after 1854, Paleontology and administrative work began to claim much of the time he would willingly have bestowed upon distinctly zoological research. His lectures on Natural History of course demanded a good deal of first-hand investigation, and not only occasional notes in his fragmentary journals, but a vast mass of drawings now preserved at South Kensington attest the amount of work he still managed to give to these subjects. But with the exception of the Hunterian Lectures of 1868, he only published one paper on Invertebrates as late as 1860; and only half a dozen, not counting the belated "Oceanic Hydrozoa," between 1856 and 1859. The essay on the Crayfish did not appear until after he had left Jermyn Street and Paleontology for South Kensington.

The "Method of Paleontology," published in 1856, was the first of a long series of papers dealing with fossil creatures, the description of which fell to him as Naturalist to the Geological Survey. By 1860 he had published twelve such papers, and by 1871 twenty-six more, or thirty-eight in sixteen years.

It was a curious irony of fate that led him into

this position.   He writes in his Autobiography that,
when Sir Henry de la Beche, the Director-General of the
Geological Survey, offered him the post Forbes vacated
of Paleontologist and Lecturer on Natural History,

> I refused the former point blank, and accepted the
> latter only provisionally, telling Sir Henry that I did
> not care for fossils, and that I should give up Natural
> History as soon as I could get a physiological post.   But
> I held the office for thirty-one years, and a large part of
> my work has been paleontological.

Yet the diversion was not without great use.   A
wide knowledge of paleontology offered a key to
many problems that were hotly debated in the years
of battle following the publication of the *Origin of
Species* in 1859, as well as providing fresh subject-
matter for the lectures in which he continued to give
the lay world the results of his thought.

On the administrative and official side he laid
before himself the organisation of the resources of
the Museum of Practical Geology as an educational
instrument.   This involved several years' work in
the arrangement of the specimens, so as to illustrate
the paleontological lectures, and the writing of
"introductions" to each section of the catalogue,
which should be a guide to the students.   The
"Method of Paleontology" mentioned above served
as the prefatory essay to the whole catalogue, and
was reprinted in 1869 by the Smithsonian Institute
of Washington under the title of *Principles and
Methods of Paleontology.*

This work led to his taking a lively interest in the organisation of museums in general, whether private, such as Sir Philip Egerton's,[1] which he visited in 1856; local, such as Warwick or Chester; or central, such as the British Museum or that at Manchester.

With regard to the British Museum, the question had arisen of removing the Natural History collections from the confined space and dusty surroundings of Great Russell Street. A first memorial on the subject had been signed, not only by many non-scientific persons, but also by a number of botanists who wished to see the British Museum Herbarium, etc., combined with the more accessible and more complete collections at Kew. Owing apparently to official opposition, the Natural History sub-committee of the British Museum Trustees advised a treatment of the Botanical Department which commended itself to none of the leading botanists. Consequently a number of botanists and zoologists took counsel together and drew up a fresh memorial from the strictly scientific point of view. Huxley and Hooker took an active part in the agitation. "It is no use," writes the former to his friend, "putting any faith in the old buffers, hardened as they are in trespasses and sin." And again :—

I see nothing for it but for you and I to constitute ourselves into a permanent "Committee of Public Safety,"

---

[1] Sir Philip de Malpas Grey Egerton, 1806-1881. His collection of fossil fishes was acquired for the Natural History Museum, South Kensington.

to watch over what is being done and take measures with
the advice of others when necessary. . . . As for ——
and *id genus omne*, I have never expected anything but
opposition from them. But I don't think it is necessary
to trouble one's head about such opposition. It may be
annoying and troublesome, but if we are beaten by it we
deserve to be. We shall have to wade through oceans
of trouble and abuse, but so long as we gain our end,
I care not a whistle whether the sweet voices of the
scientific mob are with me or against me.

According to Huxley's views a complete system
demanded a triple museum for each subject, Zoology
and Botany, since Geology was sufficiently provided
for in Jermyn Street—one typical or popular, "in
which all prominent forms or types of animals or
plants, recent or fossil, should be so displayed as to
give the public an idea of the vast extent and variety
of natural objects, to diffuse a general knowledge of
the results obtained by science in their investigation
and classification, and to serve as a general introduction to the student in Natural Science"; the second
scientific, "in which collections of all available animals
and plants and their parts, whether recent or fossil,
and in a sufficient number of specimens, should be
disposed conveniently for study, and to which should
be exclusively attached an appropriate library, or
collection of books and illustrations relating to science,
quite independent of any general library"; the third
economic, "in which economic products, whether
zoological or botanical, with illustrations of the pro-
cesses by which they are obtained and applied to use,

should be so disposed as best to assist the progress of Commerce and the Arts." It demanded further a Zoological and a Botanical Garden, where the living specimens could be studied.

Some of these institutions existed, but were not under state control. Others were already begun—*e.g.* that of Economic Zoology at South Kensington; but the value of the botanical collections was minimised by want of concentration, while as to zoology "the British Museum contains a magnificent collection of recent and fossil animals, the property of the state, but there is no room for its proper display and no accommodation for its proper study. Its official head reports directly neither to the Government nor to the governing body of the institution. . . . It is true that the people stroll through the enormous collections of the British Museum, but the sole result is that they are dazzled and confused by the multiplicity of unexplained objects, and the man of science is deprived thrice a week of the means of advancing knowledge."

The agitation of 1859-60 bore fruit in due season, and within twenty years the ideal here sketched was to a great extent realised, as any visitor to the Natural History Museum at South Kensington can see for himself.

The same principles are reiterated in his letter of January 25, 1868, to the Commissioners of the Manchester Natural History Society, who had asked his advice as to the erection of a museum. But to the

principles he adds a number of most practical suggestions as to the actual structure of the building, which
are briefly appended in abstract. The complement
to this is a letter of 1872, giving advice as to a local
museum at Chester, and one of 1859 describing the
ideal catalogue for a geological museum. (Cp. also
ii. 449.)

Jan. 25, 1868.

The Commissioners of the Manchester
        Natural History Society.

Scheme for a Museum.

Objects.—1. The public exhibition of a collection of
specimens large enough to illustrate all the most important
truths of Natural History, but not so extensive as to
weary and confuse ordinary visitors.

2. The accessibility of this collection to the public.

3. The conservation of all specimens not necessary for
the purpose defined in (1) in a place apart.

4. The accessibility of all objects contained in the
museum to the curator and to scientific students, without
interference with the public or by the public.

5. Thorough exclusion of dust and dirt from the
specimens.

6. A provision of space for workrooms, and, if need
be, lecture-rooms.

Principle.—A big hall (350 × 40 × 30) with narrower
halls on either side, lighted from the top. The central
hall for the public, the others for the curators, etc. The
walls, of arches upon piers about 15 ft. high, bearing on
girders a gallery 5 ft. wide in the public room, and 3 ft.
6 in. in the curators'.

The cases should be larger below, 5 ft. deep, and
smaller above, 2 ft. deep, with glass fronts to the public,
and doors on the curators' side.

For very large specimens—*e.g.* a whale—the case could expand into the curators' part without encroaching on the public part, so as to keep the line of windows regular.

Specimens of the Vertebrata, illustrations of Physical Geography and Stratigraphical Geology, should be placed below.

The Invertebrata, Botanical and Mineralogical specimens in the galleries.

The partition to be continued above the galleries to the roof, thus excluding all the dust raised by the public.

Space for students should be provided in the curators' rooms.

Storage should be *ample.*

A museum of this size gives twice as much area for exhibition purposes as that offered by *all* the cases in the present museum.

ATHENÆUM CLUB, *Dec.* 8, 1872.

DEAR SIR—I regret that your letter has but just come into my hands, so that my reply cannot be in time for your meeting, which, I understand you to say, was to be held yesterday.

I have no hesitation whatever in expressing the opinion that, except in the case of large and wealthy towns (and even in their case primarily), a Local Museum should be exactly what its name implies, viz. "Local"—illustrating local Geology, local Botany, local Zoology, and local Archæology.

Such a museum, if residents who are interested in these sciences take proper pains, may be brought to a great degree of perfection and be unique of its kind. It will tell both natives and strangers exactly what they want to know, and possess great scientific interest and importance. Whereas the ordinary lumber-room of clubs from New Zealand, Hindoo idols, sharks' teeth, mangy monkeys, scorpions, and conch shells—who shall describe

the weary inutility of it? It is really worse than
nothing, because it leads the unwary to look for the
objects of science elsewhere than under their noses.
What they want to know is that their "America is here,"
as Wilhelm Meister has it.—Yours faithfully,

T. H. HUXLEY.

Alfred Walker, Esq., Nant-y-Glyn, Colwyn Bay.

To THE REV. P. BRODIE OF WARWICK

JERMYN ST., *Oct.* 14, 1859.

MY DEAR MR. BRODIE—I am sorry to say that I can
as yet send you no catalogue of ours. The remodelling
of our museum is only just completed, and only the
introductory part of my catalogue is written. When it is
printed you shall have an early copy.

If I may make a suggestion, I should say that a
catalogue of your museum for popular use should
commence with a sketch of the topography and strati-
graphy of the county, put into the most intelligible
language, and illustrated by reference to mineral specimens in the cases, and to the localities where sections
showing the superposition of such and such beds are to be
seen. After that I think should come a list of the most
remarkable and interesting fossils, with reference to the
cases where they are to be seen; and under the head of
each a brief popular account of the kind of animal or
plant which the thing was when alive, its probable habits,
and its meaning and importance as a member of the great
series of successive forms of life.—Yours very faithfully,

T. H. HUXLEY.

The reorganisation of the course of studies at
Jermyn Street, fully sketched out in the 1857 notebook, involved two very serious additions to his

work over and above what was required of him by his appointment as Professor. He found his students to a great extent lacking in the knowledge of general principles necessary to the comprehension of the special work before them. To enable them to make the best use of his regular lectures, he offered them in addition a preliminary evening course of nine lectures each January, which he entitled "An Introduction to the Study of the Collection of Fossils in the Museum of Practical Geology." These lectures summed up what he afterwards named Physiography, together with a general sketch of fossils and their nature, the classification of animals and plants, their distribution at various epochs, and the principles on which they are constructed, illustrated by the examination of some animal, such as a lobster.

The regular lectures, fifty-seven in number, ran from February to April and from April to June, with fortnightly examinations during the latter period, six in number. I take the scheme from his notebook:—
"After prolegomena, the physiology and morphology of lobster and dove; then through Invertebrates, Anodon, Actinia, and Vorticella Protozoa, to Molluscan types. Insects, then Vertebrates. Supplemented Paleontologically by the demonstrations of the selected types in the cases; twelve Paleozoic, twelve Mesozoic and Cainozoic," by his assistants. "To make the course complete there should be added (1) A series of lectures on Species, practical discrimination and description, modification by conditions and distribu-

tion ; (2) Lectures on the elements of Botany and Fossil Plants."

This reorganisation of his course went hand in hand with his utilisation of the Jermyn Street Museum for paleontological teaching, and all through 1857 he was busily working at the Explanatory Catalogue.

Moreover, in 1855 he had begun at Jermyn Street his regular courses of lectures to working men— lectures which impressed those qualified to judge as surpassing even his class lectures. Year after year he gave the artisans of his best, on the principle enunciated thus early in a letter of February 27, 1855, to Dyster—

I enclose a prospectus of some People's Lectures (*Popular* Lectures I hold to be an abomination unto the Lord) I am about to give here. I want the working classes to understand that Science and her ways are great facts for them—that physical virtue is the base of all other, and that they are to be clean and temperate and all the rest—not because fellows in black with white ties tell them so, but because these are plain and patent laws of nature which they must obey "under penalties."

I am sick of the dilettante middle class, and mean to try what I can do with these hard-handed fellows who live among facts. You will be with me, I know.

And again on May 6, 1855 :—

I am glad your lectures went off so well. They were better attended than mine [the Preliminary Course], although in point of earnestness and attention my audience was all I could wish. I am now giving a course of the same kind to working men exclusively—one of what we call our series of "working men's lectures," consisting of six given in turn by each Professor. The theatre holds 600, and is crammed full.

I believe in the fustian, and can talk better to it than to any amount of gauze and Saxony; and to a fustian audience (but to that only) I would willingly give some when I come to Tenby.

The corresponding movement set going by F. D. Maurice also claimed his interest, and in 1857 he gave his first address at the Working Men's College to an audience, as he notes, of some fifty persons, including Maurice himself.

Other work of importance was connected with the Royal Institution. He had been elected to deliver the triennial course as Fullerian Professor, and for his subject in 1856-57 chose Physiology and Comparative Anatomy; in 1858, the Principles of Biology.

He was extremely glad of the additional "grist to the mill" brought in by these lectures, for it may be noted that the self-imposed burden of assisting a relative who had fallen into difficulties, and of providing for the education of her children, bore with especial weight upon the earlier years of his married life. As he wrote in 1890 :—

I have good reason to know what difference a hundred a year makes when your income is not more than four or five times that. I remember when I was candidate for the Fullerian professorship some twenty-three years ago, a friend of mine asked a wealthy manager to support me. He promised, but asked the value of the appointment, and when told, said, "Well, but what's the use of a hundred a year to him?" I suppose he paid his butler that.

A further attempt to organise scientific work throughout the country and make its results generally known, dates from this time. Huxley, Hooker, and Tyndall had discussed, early in 1858, the possibility of starting a *Scientific Review*, which should do for science what the *Quarterly* or the *Westminster* did for literature. The scheme was found not to be feasible at the time, though it was revived in another form in 1860 ; so in the meanwhile it was arranged that science should be laid before the public every fortnight, through the medium of a scientific column in the *Saturday Review*. The following letter bears on this proposal :—

April 20, 1858.

MY DEAR HOOKER—Before the dawn of the proposal for the ever-memorable though not-to-be *Scientific Review*, there had been some talk of one or two of us working the public up for science through the *Saturday Review*. Maskelyne [1] (you know him, I suppose) was the suggester of the scheme, and undertook to talk to the *Saturday* people about it.

I thought the whole affair had dropped through, but yesterday Maskelyne came to me and to Ramsay with definite propositions from the *Saturday* editor.

He undertakes to put in a scientific article in the intermediate part between Leaders and Reviews once a fortnight if we will supply him. He is not to mutilate or to alter, but to take what he gets and be thankful.

---

[1] M. H. Neville Story Maskelyne, F.R.S., Professor of Mineralogy at Oxford, 1856-1895.

The writers to select their own subjects. Now the question is, Will seven or eight of us, representing different sciences, join together and undertake to supply at least one article in three months? Once a fortnight would want a minimum of six articles in three months, so that if there were six, each man must supply one.

Sylvester is talked of for Mathematics. I am going to write to Tyndall about doing Physics. Maskelyne and perhaps Frankland will take Chemistry and Mineralogy. You and I might do Biology ; Ramsay, Geology; Smyth, Technology.

This looks to me like a very feasible plan, not asking too much of any one, and yet giving all an opportunity of saying what he has to say.

Besides this the *Saturday* would be glad to get Reviews from us.

If all those mentioned agree to join, we will meet somewhere and discuss plans.

Let me have a line to say what you think, and believe me, ever yours faithfully,   T. H. HUXLEY.

In 1858 he read three papers at the Geological and two at the Linnean ; he lectured (February 15) on Fish and Fisheries at South Kensington, and on May 21 gave a Friday evening discourse at the Royal Institution on " The Phenomena of Gemmation." He wrote an article for *Todd's Cyclopædia*, on the Tegumentary Organs, an elaborate paper, as Sir M. Foster says, on a histological theme, to which, as to others of the same class on the Teeth and the Corpuscula Tactus (*Q. J. Micr. Sci.* 1853-4), he had been " led probably by the desire, which only gradually and through lack of fulfilment left him, to become a physiologist rather than a naturalist."

No less important was his more general work for science. Physiological study in England at this time was dominated by transcendental notions. To put first principles on a sound experimental basis was the aim of the new leaders of scientific thought. To this end Huxley made two contributions in the fifties— one on the general subject of the cell theory, the other on the particular question of the development of the skull. "In a striking 'Review of the Cell Theory,'" says Sir M. Foster, "which appeared in the *British and Foreign Medical Review* in 1853, a paper which more than one young physiologist at the time read with delight, and which even to-day may be studied with no little profit, he, in this subject as in others, drove the sword of rational inquiry through the heart of conceptions, metaphysical and transcendental, but dominant."

Of this article Professor E. Ray Lankester also writes :—

. . . Indeed it is a fundamenal study in morphology. The extreme interest and importance of the views put forward in that article may be judged of by the fact that although it is forty years since it was published, and although our knowledge of cell structure has made immense progress during those forty years, yet the main contention of that article, viz. that cells are not the cause but the result of organisation—in fact, are, as he says, to the tide of life what the line of shells and weeds on the sea-shore is to the tide of the living sea—is even now being reasserted, and in a slightly modified form is by very many cytologists admitted as having more truth in it than the opposed view and its later outcomes, to the effect that the cell is the unit of life in which and through which alone living matter manifests its activities.

The second was his Croonian Lecture of 1858, "On the Theory of the Vertebrate Skull," in which he demonstrated from the embryological researches of Rathke and others, that after the first step the whole course of development in the segments of the skull proceeded on different lines from that of the vertebral column ; and that Oken's imaginative theory of the skull as modified vertebræ, logically complete down to a strict parallel between the subsidiary head-bones and the limbs attached to the spine, outran the facts of a definite structure common to all vertebrates which he had observed.[1]

With the demolition of Oken's theory fell the superstructure raised by its chief supporter, Owen, "archetype" and all.

[1] "Following up Rathke, he strove to substitute for the then dominant fantastic doctrines of the homologies of the cranial elements advocated by Owen, sounder views based on embryological evidence. He exposed the futility of attempting to regard the skull as a series of segments, in each of which might be recognised all the several parts of a vertebra, and pointed out the errors of trusting to superficial resemblances of shape and position. He showed, by the history of the development of each, that, though both skull and vertebral column are segmented, the one and the other, after an early stage, are fashioned on lines so different as to exclude all possibility of regarding the detailed features of each as mere modifications of a type repeated along the axis of the body. 'The spinal column and the skull start from the same primitive condition, whence they immediately begin to diverge.' 'It may be true to say that there is a primitive identity of structure between the spinal or vertebral column and the skull ; but it is no more true that the adult skull is a modified vertebral column than it would be to affirm that the vertebral column is modified skull.' This lecture marked an epoch in England in vertebrate morphology,

It was undoubtedly a bold step to challenge thus openly the man who was acknowledged as the autocrat of science in Britain. Moreover, though he had long felt that on his own subjects he was Owen's master, to begin a controversy was contrary to his deliberate practice. But now he had the choice of submitting to arbitrary dictation or securing himself from further aggressions by dealing a blow which would weaken the authority of the aggressor. For the growing antagonism between him and Owen had come to a head early in the preceding year, when the latter, taking advantage of the permission to use the lecture-theatre at Jermyn Street for the delivery of a paleontological course, unwarrantably assumed the title of Professor of Paleontology at the School of Mines, to the obvious detriment of Huxley's position there. His explanations not satisfying the council of the School of Mines, Huxley broke off all personal intercourse with him.

and the views enunciated in it carried forward, if somewhat modified, as they have been, not only by Huxley's subsequent researches and by those of his disciples, but especially by the splendid work of Gegenbaur, are still, in the main, the views of the anatomists of to-day."—Sir M. FOSTER, Royal Society Obituary Notice of T. H. Huxley.

CHAPTER XI

1857–1858

THROUGHOUT this period his health was greatly tried by the strain of his work and life in town. Headache! headache! is his repeated note in the early part of 1857, and in 1858 we find such entries as:—
"Feb. 11.—Used up. Hypochondriacal and bedevilled." "Ditto 12." "13.—Not good for much." "21.—Toothache, incapable all day." And again:—
"March 30.—Voiceless." "31.—Missed lecture." And, "April 1.—Unable to go out." He would come in thoroughly used up after lecturing twice on the same day, as frequently happened, and lie wearily on one sofa; while his wife, whose health was wretched, matched him on the other. Yet he would go down to a lecture feeling utterly unable to deliver it, and, once started, would carry it through successfully—at what cost of nervous energy was known only to those two at home.

But there was another branch of work, that for the Geological Survey, which occasionally took him out of London, and the open-air occupation and

tramping from place to place did him no little good.
Thus, through the greater part of September and
October 1856 he ranged the coasts of the Bristol
Channel from Weston to Clovelly, and from Tenby
to Swansea, preparing a "Report on the Recent
Changes of Level in the Bristol Channel." "You
can't think," he writes from Braunton on October 3,
"how well I am, so long as I walk eight or ten miles
a day and don't work too much, but I find fifteen or
sixteen miles my limit for comfort."

For many years after this his favourite mode of
recruiting from the results of a spell of overwork
was to take a short walking tour with a friend.   In
April 1857 he is off for a week to Cromer; in 1860
he goes with Busk and Hooker for Christmas week
to Snowdon; another time he is manœuvred off by
his wife and friends to Switzerland with Tyndall.

In Switzerland he spent his summer holidays both
in 1856 and 1857, in the latter year examining the
glaciers with Tyndall scientifically, as well as seeking
pleasure by the ascent of Mont Blanc.   As fruits of
this excursion were published late in the same year,
his "Letter to Mr. Tyndall on the Structure of
Glacier Ice" (*Phil. Mag.* xiv. 1857), and the paper
in the *Philosophical Transactions of the Royal Society*,
which appeared—much against his will—in the joint
names of himself and Tyndall.   Of these he wrote in
1893 in answer to an inquiry on the subject :—

By the Observations on Glaciers I imagine you refer
to a short paper published in *Phil. Mag.* that embodied

results of a little bit of work of my own. The Glacier paper in the *Phil. Trans.* is essentially and in all respects Professor Tyndall's. He took up glacier work in consequence of a conversation at my table, and we went out to Switzerland together, and of course talked over the matter a good deal. However, except for my friend's insistence, I should not have allowed my name to appear as joint author, and I doubt whether I ought to have yielded. But he is a masterful man and over-generous.

And in a letter to Hooker he writes :—

By the way, you really must not associate me with Tyndall and talk about *our* theory. My sole merit in the matter (and for that I do take some credit) is to have set him at work at it, for the only suggestion I made, viz. that the veined structure was analogous to his artificial cleavage phenomena, has turned out to be quite wrong.

Tyndall fairly *made* me put my name to that paper, and would have had it first if I would have let him, but if people go on ascribing to me any share in his admirable work I shall have to make a public protest. All I am content to share is the row, if there is to be one.

The following letters to Hooker and Tyndall touch upon his Swiss trips of 1856 and 1857 :—

BERNE, *Sept.* 3, 1856.

I send you a line hence, having forgotten to write from Interlaken, whence we departed this morning.

The Weissthor expedition was the most successful thing you can imagine. We reached the Riffelberg in $11\frac{1}{2}$ hours, the first six being the hardest work I ever had in my life in the climbing way, and the last five carrying us through the most glorious sight I ever witnessed. During the latter part of the day there was not a cloud on the whole Monte Rosa range, so you may imagine what the Matterhorn and the rest of them looked like from the wide plain of névé just below the Weissthor. It was quite a new sensation, and I would not have missed it for any amount; and besides this I had an opportunity of examining the névé at a very great height. A regularly stratified section, several hundred feet high, was exposed on the Cima di Jazi, and I was convinced that the Weissthor would be a capital spot for making observations on the névé and on other correlative matters. There are no difficulties in the way of getting up to it from the Zermatt side, tough job as it is from Macugnaga, and we might readily rig a tent under shelter of the ridge. That would lick old Saussure into fits. All the Zermatt guides put the S. Theodul pass far beneath the Weissthor in point of difficulty ; and you may tell Mrs. Hooker that they think the S. Theodul easier than the Monte Moro. The best of the joke was that I lost my way in coming down the Riffelberg to Zermatt the same evening, so that altogether I had a long day of it. The next day I walked from Zermatt to Visp (recovering Baedeker by the way), but my shoes were so knocked to pieces that I got a blister on my heel. Next day voiture to Susten, and then over Gemmi to Kandersteg, and on Thursday my foot was so queer I was glad to get a retour to Interlaken. I found most interesting and complete evidences of old moraine deposits all the way down the Leuk valley into the Rhone valley, and I believe those little hills beyond Susten are old terminal moraines too. On the other side I followed moraines down to Frutigen, and great masses of glacial gravel with boulders, nearly to the Lake of Thun.

My wife is better, but anything but strong.

CHAMOUNIX, *Aug.* 16, 1857.

My wife sends me intelligence of the good news you were so kind as to communicate to her. I need not tell you how rejoiced I am that everything has gone on well, and that your wife is safe and well. Offer her my warmest congratulations and good wishes. I have made one matrimonial engagement for Noel already, otherwise I would bespeak the hand of the young lady for him.

It has been raining cats and dogs these two days, so that we have been unable to return to our headquarters at the Montanvert which we left on Wednesday for the purpose of going up Mont Blanc. Tyndall (who has become one of the most active and daring mountaineers you ever saw—so that we have christened him "cat"; and our guide said the other day, " Il va plus fort qu'un mouton. Il faut lui mettre une sonnette ") had set his heart on the performance of this feat (of course with purely scientific objects), and had equally made up his mind not to pay five-and-twenty pounds for the gratifica-tion. So we had one guide and took two porters in addition as far as the Grands Mulets. He is writing to you, and will tell you himself what happened to those who reached the top—to wit, himself, Hirst, and the guide. I found that three days in Switzerland had not given me my Swiss legs, and consequently I remained at the Grands Mulets, all alone in my glory, and for some eight hours in a great state of anxiety, for the three did not return for about that period after they were due.

I was there on a pinnacle like St. Simon Stylites, and nearly as dirty as that worthy saint must have been, but without any of his other claims to angelic assistance, so that I really did not see, if they had fallen into a crevasse, how I was to help either them or myself. They came back at last, just as it was growing dusk, to my inexpressible relief, and the next day we came down here—such a set of dirty, sunburnt, snow-blind wretches as you never saw.

We heartily wished you were with us. What we shall do next I neither know nor care, as I have placed

myself entirely under Commodore Tyndall's orders ; but
I suppose we shall be three or four days more at the
Montanvert, and then make the tour of Mont Blanc. I
have tied up six pounds in one end of my purse, and
when I have no more than that I shall come back.
Altogether I don't feel in the least like the father of
a family ; no more would you if you were here. The
habit of carrying a pack, I suppose, makes the "quiver
full of arrows" feel light.

115 ESPLANADE, DEAL,
Sept. 3, 1857.

MY DEAR TYNDALL—I don't consider myself returned
until next Wednesday, when the establishment of No.
14 will reopen on its accustomed scale of magnificence,
but I don't mind letting you know I am in the flesh and
safe back.

The tour round Mont Blanc was a decided success ;
in fact, I had only to regret you were not with me. The
grand glacier of the Allée Blanche and the view of Mont
Blanc from the valley of Aosta were alone worth all the
trouble. I had only one wet day, and that I spent on
the Brenoa Glacier ; for, in spite of all good resolutions to
the contrary, I cannot resist poking into the glaciers
whenever I have a chance. You will be interested in
my results, which we shall soon, I hope, talk on together
at length.

As I suspected, Forbes [1] has made a most egregious
blunder. What he speaks of and figures as the " structure "
of the Brenoa is nothing but a peculiar arrangement of
*entirely superficial dirt bands, dependent on the structure, but
not it.* The true structure is singularly beautiful and
well marked in the Brenoa, the blue veins being very
close set, and of course wholly invisible from a distance of

---

[1] James David Forbes, 1809-1868, Professor of Natural Philosophy at Edinburgh, 1833 ; Principal of the United College,
St. Andrews, 1859. Since 1841 he had been investigating the
nature of glaciers. See p. 216.

a hundred yards, which is less than that of the spot whence Forbes' view of the (supposed) structure is taken.

I saw another wonderful thing in La Brenoa. About the middle of its length there is a step like this of about 20 or 30 feet in height. In the lower part (B) the structural planes are vertical; in the upper (A) they dip at a considerable angle. I thought I had found a case of unconformability, indicating a slip of one portion of the glacier over another, but when I came to examine the intermediate region (X) carefully, I found the structural planes at every intermediate angle, and consequently a perfect transition from the one to the other.

I returned by Aosta, the Great St. Bernard, and the Col de Balme. Old Simond was quite affectionate in his discourse about you, and seemed quite unhappy because you would not borrow his money. He had received your remittance, and asked me to tell you so. He was distressed at having forgotten to get a certificate from you, so I said in mine that I was quite sure you were well satisfied with him.

On our journey he displayed his characteristic qualities, *Je ne sais pas* being the usual answer to any topographical inquiries, with a total absence of nerve, and a general conviction that distances were very great and that the weather would be bad. However, we got on very well, and I was sorry to part with him.

I came home by way of Neuchatel, paying a visit to the Pierre à Bôt, which I have long wished to see. My financial calculations were perfect in theory, but nearly broke down in practice, inasmuch as I was twice obliged to travel first-class when I calculated on second. The result was that my personal expenses between Paris and London amounted to 1.50!! and I arrived at my own house hungry and with a remainder of a few centimes. I should think that your fate must have been similar.

Many thanks for writing to my wife. She sends her kindest remembrances to you.—Ever yours,   T. H. H.

The year 1857 was the last in which Huxley
apparently had time to go so far in journal-writing as
to draw up a balance-sheet at the year's end of work
done and work undone. Though he finds " as usual
a lamentable difference between agenda and acta;
many things proposed to be done not done, and many
things not thought of finished," still there is enough
noted to satisfy most energetic people. Mention has
already been made of his lectures — sixty-six at
Jermyn Street, twelve Fullerian, and as many more
to prepare for the next year's course; seven to working men, and one at the Royal Institution, together
with the rearrangement of specimens at the Jermyn
Street Museum, and the preparation of the Explanatory Catalogue, which this year was published to the
extent of the Introduction and the Tertiary collections.
To these may be added examinations at the London
University, where he had succeeded Dr. Carpenter as
examiner in Physiology and Comparative Anatomy
in 1856, reviews, translations, a report on Deep Sea
Soundings, and ten scientific memoirs.

The most important of the unfinished work consists of the long-delayed *Oceanic Hydrozoa*, the *Manual
of Comparative Anatomy*, and a report on Fisheries.
The rest of the unfinished programme shows the
usual commixture of technical studies in anatomy and
paleontology, with essays on the philosophical and
educational bearings of his work. On the one hand
are memoirs of Daphnia, Nautilus, and the Herring,
the affinities of the Paleozoic Crustacea, the Ascidian

Catalogue and Positive Histology; on the other, the Literature of the Drift, a review of the present state of philosophical anatomy, and a scheme for arranging the Explanatory Catalogue to serve as an introductory text-book to the Jermyn Street lectures and the paleontological demonstrations. Here, too, would fall a proposed "Letter on the Study of Comparative Anatomy," to do for those subjects what Henslow had done in his "Letter" for Botany.

In addition to the fact of his being forced to take up Paleontology, it was perhaps the philosophic breadth of view with which he regarded his subject at any time, and the desire of getting to the bottom of each subsidiary problem arising from it, that made him for many years seem constantly to spring aside from his own subject, to fly off at a tangent from the line in which he was assured of unrivalled success did he but devote to it his undivided powers. But he was prepared to endure the charge of desultoriness with equanimity. In part, he was still studying the whole field of biological science before he would claim to be a master in one department; in part, he could not yet tell to what post he might succeed when he left—as he fully expected to leave—the professorship at Jermyn Street.

One characteristic of his early papers should not pass unnoticed. This was his familiarity with the best that had been written on his subjects abroad as well as in England. Thoroughness in this respect was rendered easier by the fact that he read French and German with almost as much facility as his mother tongue. "It is true, of course, that scientific men read French and German before the time of Huxley; but the deliberate consultation of all the authorities available has been maintained in historical succession since Huxley's earliest papers, and was absent in the papers of his early contemporaries."[1]

About this time his activity in several branches of science began to find recognition from scientific societies at home and abroad. In 1857 he was elected honorary member of the Microscopical Society of Giessen; and in the same year, of a more important body, the Academy of Breslau (Imperialis Academia Cæsariana Naturæ Curiosorum). He writes to Hooker:—

14 WAVERLEY PLACE, *April* 3, 1857.

Having subsided from standing upon my head—which was the immediate causation of your correspondence about the co-extension Imp. Acad. Cæs. Nat. Cur. (don't I know their thundering long title well!)—I have to say that I was born on the 4th of May of the year 1825, whereby I have now more or less mis-spent thirty-one years and a bittock, nigh on thirty-two.

Furthermore, my *locus natalis* is Ealing, in the county of Middlesex. Upon my word, it is very obliging of the "curious naturals," and I must say wholly surprising and unexpected.

I shall hold up my head immensely to-morrow when (blessed be the Lord) I give my last Fullerian.

Among other things, I am going to take Cuvier's crack case of the 'Possum of Montmartre as an illustration of *my* views.

---

[1] P. Chalmers Mitchell in *Natural Science*, August 1895.

I wondered what had become of you, but the people have come talking about me this last lecture or two, so I supposed you had erupted to Kew.

My glacier article is out; tell me what you think of it some day.

I wrote a civil note to Forbes[1] yesterday, charging myself with my crime, and I hope that is the end of the business.

My wife is mending slowly, and if she were here would desire to be remembered to you.

In December 1858 he became a Fellow of the Linnean, and the following month not only Fellow but Secretary of the Geological Society.

In 1858 also he was elected to the Athenæum Club under Rule 2, which provides that the committee shall yearly elect a limited number of persons distinguished in art, science, or letters. His proposer was Sir Roderick Murchison, who wrote :—

ATHENÆUM, *Jan.* 26.

MY DEAR HUXLEY—I had a success as to you that I never had or heard of before. Nineteen persons voted, and of these eighteen voted for you and no one against you. You, of course, came in at the head of the poll ; no other having, *i.e.* Cobden, more than eleven.—Yours well satisfied,   ROD. I. MURCHISON.

From this time forth he corresponded with many foreign men of science ; in these years particularly with Victor Carus, Lacaze Duthiers, Kölliker, and de Quatrefages, in reference to their common interest in the study of the invertebrates.

---

[1] Principal James Forbes, with whose theory of glaciers Huxley and Tyndall disagreed.

At home, the year 1857 opened very brightly for Huxley with the birth of his first child, a son, on the eve of the New Year. A Christmas child, the boy was named Noel, and lived four happy years to be the very sunshine of home, the object of passionate devotion, whose sudden loss struck deeper and more ineffaceably than any other blow that befell Huxley during all his life.

As he sat alone that December night, in the little room that was his study in the house in Waverley Place, waiting for the event that was to bring him so much happiness and so much sorrow, he made a last entry in his journal full of hope and resolution. In the blank space below follows a note of four years later, when "the ground seemed cut from under his feet," yet written with restraint and without bitterness.

December 31, 1856 . . . 1856-7-8 must still be "Lehrjahre" to complete training in principles of Histology, Morphology, Physiology, Zoology, and Geology by *Monographic Work* in each Department. 1860 will then see me well grounded and ready for any special pursuits in either of these branches.

It is impossible to map out beforehand how this must be done. I must seize opportunities as they come, at the risk of the reputation of desultoriness.

In 1860 I may fairly look forward to fifteen or twenty years "Meisterjahre," and with the comprehensive views my training will have given me, I think it will be possible in that time to give a new and healthier direction to all Biological Science.

To smite all humbugs, however big; to give a nobler tone to science; to set an example of abstinence from

petty personal controversies, and of toleration for everything but lying ; to be indifferent as to whether the work
is recognised as mine or not, so long as it is done :—are
these my aims ? 1860 will show.

> Willst du dir ein hübsch Leben zimmern,
> Musst dich ans Vergangene nicht bekümmern ;
> Und wäre dir auch was Verloren,
> Musst immer thun wie neugeboren.
> Was jeder Tag will, sollst du fragen ;
> Was jeder Tag will, wird er sagen.
> Musst dich an eigenem Thun ergötzen ;
> Was andere thun, das wirst du schätzen.
> Besonders keinen Menschen hassen
> Und das Übrige Gott überlassen.[1]

Half-past ten at night.
     Waiting for my child. I seem to fancy it the pledge
that all these things shall be.
     Born five minutes before twelve. Thank God. New
Year's Day, 1857.

                                        September 20, 1860.

     And the same child, our Noel, our first-born, after
being for nearly four years our delight and our joy, was
carried off by scarlet fever in forty-eight hours. This day
week he and I had a great romp together. On Friday
his restless head, with its bright blue eyes and tangled
golden hair, tossed all day upon his pillow. On Saturday
night the fifteenth, I carried him here into my study, and

> [1] Wilt shape a noble life ? Then cast
> No backward glances to the past.
> And what if something still be lost ?
> Act as new-born in all thou dost.
> What each day wills, that shalt thou ask ;
> Each day will tell its proper task ;
> What others do, that shalt thou prize,
> In thine own work thy guerdon lies.
> This above all : hate none. The rest—
> Leave it to God. He knoweth best.

laid his cold still body here where I write.   Here too on Sunday night came his mother and I to that holy leave-taking.

My boy is gone, but in a higher and a better sense than was in my mind when I wrote four years ago what stands above—I feel that my fancy has been fulfilled.   I say heartily and without bitterness—Amen, so let it be.

CHAPTER XII

1859–1860

THE programme laid down in 1857 was steadily carried out through a great part of 1859. Huxley published nine monographs, chiefly on fossil Reptilia, in the proceedings of the Geological Society and of the Geological Survey, one on the Armour of Crocodiles at the Linnean, and "Observations on the Development of some Parts of the Skeleton of Fishes," in the *Journal of Microscopical Science.*

Among the former was a paper on Stagonolepis, a creature from the Elgin beds, which had previously been ranked among the fishes. From some new remains, which he worked out of the stone with his own hands, Huxley made out that this was a reptile closely allied to the Crocodiles; and from this and the affinities of another fossil, Hyperodapedon, from neighbouring beds, determined the geological age to which the Elgin beds belonged. A good deal turned upon the nature of the scales from the back and belly of this animal, and a careful comparison with the scales of modern crocodiles—a subject till then little investigated—led to the paper at the Linnean already mentioned.

The paper on fish-development was mainly based upon dissections of the young of the stickleback. Fishes had been divided into two classes according as their tails are developed evenly on either side of the line of the spine, which was supposed to continue straight through the centre of the tail (homocercal), or lopsided, with one tail fin larger than the other (heterocercal). This investigation showed that the apparently even development was only an extreme case of lopsidedness, the continuation of the "chorda," which gives rise to the spine, being at the top of the upper fin, and both fins being developed on the same side of it. Lopsidedness as such, therefore, was not to be regarded as an embryological character in ancient fishes; what might be regarded as such was the absence of a bony sheath to the end of the "chorda" found in the more developed fishes. Further traces of this bony structure were shown to exist, among other piscine resemblances, in the Amphibia. Finally the embryological facts now observed in the development of the bones of the skull were of great importance, "as they enable us to understand, on the one hand, the different modifications of the palato-suspensorial apparatus in fishes, and on the other hand the relations of the components of this apparatus to the corresponding parts in other *Vertebrata*," fishes, reptiles, and mammals presenting a well-marked series of gradations in respect to this point.

This part of the paper had grown out of the investigations begun for the essay on the Vertebrate Skull,[1] just as that on Jacare and Caiman from inquiry into the scales of Stagonolepis.

Thus he was still able to devote most of his time to original research. But though in his letter of March 27, 1855, below, he says, "I never write for the Reviews now, as original work is much more to my taste," it appears from jottings in his 1859 notebook, such as "Whewell's *History of Scientific Ideas*, as a Peg on which to hang Cuvier article," that he again found it necessary to supplement his income by writing. He was still examiner at London University, and delivered six lectures on Animal Motion at the London Institution and another at Warwick. This lecture he had offered to give at the Warwick Museum as some recognition of the willing help he had received from the assistants when he came down to examine certain fossils there. On the way he visited Dr. Rolleston[2] at Oxford. The knowledge of Oxford life gained from this and a later visit led him to write :—

The more I see of the place the more glad I am that I elected to stay in London. I see much to admire and like ; but I am more and more convinced that it would not suit me as a residence.

Two more important points remain to be mentioned

---

[1] See p. 203.
[2] George Rolleston, 1829-1881, Regius Professor of Physic at Oxford, 1857 ; Linacre Professor of Comparative Anatomy, 1860.

among the occupations of the year. In January
Huxley was elected Secretary of the Geological
Society, and with this office began a form of adminis-
trative work in the scientific world which ceased only
with his resignation of the Presidency of the Royal
Society in 1885.

Part of the summer Huxley spent in the North.
On August 3 he went to Lamlash Bay in Arran.
Here Dr. Carpenter had, in 1855, discovered a convenient cottage on Holy Island—the only one, indeed,
on the island—well suited for naturalists; the bay
was calm and suitable both for the dredge and for
keeping up a vivarium. He proposed that either the
Survey should rent the whole island at a cost of some
£50, or, failing this, that he would take the cottage
himself, if Huxley would join him for two or three
seasons and share the expense. Huxley laid the plan
before Sir R. Murchison, the head of the Survey,
who consented to try the plan for a course of years,
during three months in each year. "But," he added,
"keep it experimental; for there are no *useful*
fisheries such as delight Lord Stanley." Here, then,
with an ascent of Goatfell for variety on the 21st, a
month was passed in trawling, and experiments on
the spawning of the herring appear to have been
continued for him during the winter in Bute.

On the 29th Huxley left Lamlash for a trip
through central and southern Scotland, continuing
his geological work for the Survey; and wound up
by attending the meeting of the British Association

at Aberdeen, leaving his wife and the three children
at Aberdour, on the Fifeshire coast.

From Aberdeen, where Prince Albert was President
of the Association, Huxley writes on September
15 :—

Owen's brief address on giving up the presidential
chair was exceedingly good. . . . I shall be worked like
a horse here. There are all sorts of new materials from
Elgin, besides other things, and I daresay I shall have to
speak frequently. In point of attendance and money
this is the best meeting the Association ever had. In
point of science, we shall see. . . . Tyndall has accepted
the Physical chair with us, at which I am greatly delighted.

In this connection the following letter to Tyndall
is interesting :—

ABERDOUR, FIFE, N.B.
September 5, 1859.

MY DEAR TYNDALL—I met Faraday on Loch Lomond
yesterday, and learned from him that you had returned,
whereby you are a great sinner for not having written to
me. Faraday told me you were all sound, wind and
limb, and had carried out your object, which was good
to hear.

Have you had any letter from Sir Roderick ? If not,
pray call in Jermyn Street and see Reeks [1] as soon as
possible. The thing I have been hoping for for years past has
come about,—Stokes having resigned the Physical Chair
in our place, in consequence of his appointment to the
Cambridge University Commission. This unfortunately
occurred only after our last meeting for the session, and

---

[1] Mr. Trenham Reeks, who died in 1879, was Registrar of the
School of Mines, and Curator and Librarian of the Museum of
Practical Geology.

after I had left town, but Reeks wrote to me about it at once. I replied as soon as I received his letter, and told him that I would take upon myself the responsibility of saying that you would accept the chair if it were offered you. I thought I was justified in this by various conversations we have had; and, at any rate, I felt sure that it was better that I should get into a mess than that you should lose the chance.

I know that Sir Roderick has written to you, but I imagine the letter has gone to Chamounix, so pray put yourself into communication with Reeks at once.

You know very well that the having you with us at Jermyn Street is a project that has long been dear to my heart, partly on your own account, but largely for the interest of the school. I earnestly hope that there is no impediment in the way of your coming to us. How I am minded towards you, you ought to know by this time; but I can assure you that all the rest of us will receive you with open arms. Of that I am quite sure.

Let me have a line to know your determination. I am on tenterhooks till the thing is settled.

Can't you come up this way as you go to Aberdeen? —Ever yours faithfully,   T. H. HUXLEY.

P.S.—I thought I might mention the Jermyn Street matter to Faraday privately, and did so. He seemed pleased that the offer had been made.

The acceptance of the lectureship at the School of Mines brought Tyndall into the closest contact with Huxley for the next nine years, until he resigned his lectureship in 1868 on succeeding Faraday as superintendent of the Royal Institution.

On September 17 he writes :—

Yesterday Owen and I foregathered in Section D. He read a very good and important paper, and I got up afterwards and spoke exactly as I thought about it, and praising many parts of it strongly. In his reply he was unco civil and complimentary, so that the people who had come in hopes of a row were (as I intended they should be) disappointed.

A number of miscellaneous letters of this period are here grouped together.

14 WAVERLEY PLACE,
January 30, 1858.

MY DEAR HOOKER— . . . I wish you wouldn't be apologetic about criticism from people who have a right to criticise. I always look upon any criticism as a compliment, not but what the old Adam in T. H. H. *will* arise and fight vigorously against all impugnment, and irrespective of all odds in the way of authority, but that is the way of the beast.

Why I value your and Tyndall's and Darwin's friendship so much is, among other things, that you all pitch into me when necessary. You may depend upon it, however blue I may look when in the wrong, it's wrath with myself and nobody else.

To His Sister

THE GOVERNMENT SCHOOL OF MINES, JERMYN ST.,
March 27, 1858.

MY DEAREST LIZZIE—It is a month since your very welcome letter reached me. I had every inclination and every intention to answer it at once, but the wear and tear of incessant occupation (for your letter arrived in the midst of my busiest time) has, I will not say deprived me of the leisure, but of that tone of mind which one wants for writing a long letter. I fully understand—no one should be better able to comprehend—how the same causes may operate on you, but do not be silent so long again; it is bad for both of us. I have loved but few people in my life, and am not likely to care for àny more unless it be my children. I desire therefore rather to knit more firmly than to loosen the old ties, and of these which is older or stronger than ours ? Don't let us drift asunder again.

Your letter came just after the birth of my second child, a little girl. I registered her to-day in the style and title of Jessie Oriana Huxley. The second name is a family name of my wife's and not, as you might suppose, taken from Tennyson. You will know why my wife and I chose the first. We could not make you a godmother, as my wife's mother is one, and a friend of ours had long since applied for the other vacancy, but perhaps this is a better tie than that meaningless formality. My little son is fifteen months old ; a fair-haired, blue-eyed, stout little Trojan, very like his mother. He looks out on the world with bold confident eyes and open brow, as if he were its master. We shall try to make him a better man than his father. As for the little one, I am told she is pretty, and slavishly admit the fact in the presence of mother and nurse, but between ourselves I don't see it. To my carnal eyes her nose is the image of mine, and you know what that means. For though wandering up and down the world and work have begun to sow a little silver in my hair, they have by no means softened the outlines of that remarkable feature.

You want to know what I am and where I am—well, here's a list of titles. T. H. H., Professor of Natural History, Government School of Mines, Jermyn Street ; Naturalist to the Geological Survey ; Curator of the Paleontological collections (*non-official* maid-of-all-work in Natural Science to the Government); Examiner in Physiology and Comparative Anatomy to the University of London ; Fullerian Professor of Physiology to the Royal Institution (but that's just over); F.R.S., F.G.S., etc. Member of a lot of Societies and Clubs, all of

which cost him a mint of money. Considered a rising
man and not a bad fellow by his friends—*per contra*
greatly over-estimated and a bitter savage critic by his
enemies. Perhaps they are both right. I have a high
standard of excellence and am no respecter of persons,
and I am afraid I show the latter peculiarity rather
too much. An internecine feud rages between Owen
and myself (more's the pity) partly on this account,
partly from other causes.

This is the account any third person would give you
of what I am and of what I am doing. He would
probably add that I was very ambitious and desirous
of occupying a high place in the world's estimation.
Therein, however, he would be mistaken. An income
sufficient to place me above care and anxiety, and free
scope to work, are the only things I have ever wished
for or striven for. But one is obliged to toil long and
hard for these, and it is only now that they are coming
within my grasp. I gave up the idea of going to
Edinburgh because I doubted whether leaving London
was wise. Recently I have been tempted to put up
for a good physiological chair which is to be established
at Oxford ; but the Government propose to improve my
position at the School of Mines, and there is every
probability that I shall now permanently remain in
London. Indeed, it is high time that I should settle
down to one line of work. Hitherto, as you see by the
somewhat varied list of my duties, etc., above, I have
been ranging over different parts of a very wide field.
But this apparent desultoriness has been necessary, for I
knew not for what branch of science I should eventually
have to declare myself. There are very few appointments
open to men of science in this country, and one must
take what one can get and be thankful.

My health was very bad some years ago, and I had
great fear of becoming a confirmed dyspeptic, but thanks
to the pedestrian tours in the Alps I have taken for the

past two years, I am wonderfully better this session, and feel capable of any amount of work. It was in the course of one of these trips that I went, as you have rightly heard, half way up Mont Blanc. But I was not in training and stuck at the Grands Mulets, while my three companions went on. I spent seventeen hours alone on that grand pinnacle, the latter part of the time in great anxiety, for I feared my friends were lost; and as I had no guide my own neck would have been in considerable jeopardy in endeavouring to return amidst the maze of crevasses of the Glacier des Bois. But it was glorious weather and the grandest scenery in the world. In the previous year I saw much of the Bernese and Monte Rosa country, journeying with a great friend of mine well known as a natural philosopher, Tyndall, and partly seeking health and partly exploring the glaciers. You will find an article of mine on that subject in the *Westminster Review* for 1857.

I used at one time to write a good deal for that Review, principally the Quarterly notice of scientific books. But I never write for the Reviews now, as original work is much more to my taste. The articles you refer to are not mine, as, indeed, you rightly divined. The only considerable book I have translated is Kölliker's *Histology*—in conjunction with Mr. Busk, an old friend of mine. All translation and article writing is weary work, and I never do it except for filthy lucre. Lecturing I do not like much better; though one way or another I have to give about sixty or seventy a year.

Now then, I think that is enough about my "Ich." You shall have a photographic image of him and my wife and child as soon as I can find time to have them done. . . .

1 Eldon Place, Broadstairs,
Sept. 5, 1858.

MY DEAR HOOKER—I am glad Mrs. Hooker has found rest for the sole of her foot. I returned her Tyndall's letter yesterday.

Wallace's impetus seems to have set Darwin going in earnest, and I am rejoiced to hear we shall learn his views in full, at last.   I look forward to a great revolution being effected.   Depend upon it, in natural history, as in everything else, when the English mind fully determines to work a thing out, it will do it better than any other.

I firmly believe in the advent of an English epoch in science and art, which will lick the Augustan (which, by the bye, had neither science nor art in our sense, but you know what I mean) into fits.   So hooray, in the first place, for the Genera plantarum.   I can quite understand the need of a new one, and I am right glad you have undertaken it.   It seems to me to be in all respects the sort of work for you, and exactly adapted to your environment at Kew.   I remember you mentioned to me some time ago that you were thinking of it.

I wish I could even hope that such a thing would be even attempted in the course of this generation for animals.

But with animal morphology in the state in which it is now, we have no terminology that will stand, and consequently concise and comparable definitions are in many cases impossible.

If old Dom. Gray [1] were but an intelligent activity instead of being a sort of zoological whirlwind, what a deal he might do.   And I am hopeless of Owen's comprehending what classification means since the publication of the wonderful scheme which adorns the last edition of his lectures.

As you say, I have found this a great place for "work of price."   I have finished the "Oceanic Hydrozoa" all but the bookwork, for which I must have access to the B.M. Library—but another week will do him.   My notes are from eight to twelve years old, and really I often have felt like the editor of somebody else's posthumous work.

---

[1] John Edward Gray, 1800-1875, appointed Keeper of the Zoological Collections in the British Museum in 1840.

Just now I am busy over the "Croonian," which must be done before I return. I have been pulling at all the arguments as a spider does at his threads, and I think they are all strong. If so the thing will do some good.

I am perplexed about the N.H. Collections. The best thing, I firmly believe, would be for the Economic Zoology and a set of well-selected types to go to Kensington, but I should be sorry to see the scientific collection placed under any such auspices as those which govern the "Bilers." I don't believe the clay soil of the Regent's Park would matter a fraction—and to have a grand scientific zoological and paleontological collection for working purposes close to the Gardens where the living beasts are, would be a grand thing. I should not wonder if the affair is greatly discussed at the B.A. at Leeds, and then, perhaps, light will arise.

Have you seen that madcap Tyndall's letter in the *Times?* He'll break his blessed neck some day, and that will be a great hole in the efficiency of my scientific young England. We mean to return next Saturday, and somewhere about the 16th or 17th I shall go down to York, where I want to study Plesiosaurs. I shall return after the British Association. The interesting question arises, Shall I have a row with the great O. there? What a capital title that is they give him of the *British* Cuvier. He stands in exactly the same relation to the French as British brandy to cognac.—Ever yours faithfully,

T. H. HUXLEY.

Am I to send the *Gardener's Chronicle* on, and where? please. I have mislaid the address.

JERMYN STREET, *Oct.* 25, 1858.

MY DEAR SPENCER—I read your article on the "Archetype" the other day with great delight, particularly the phrase which puts the Owenian and Cummingian interpolations on the same footing. It is rayther strong, but quite just.

I do not remember a word to object to, but I think I
could have strengthened your argument in one or two
places.   Having eaten the food, will you let me have
back the dish ?   I am winding up the "Croonian," and
want *L'Archetype* to refer to.   So if you can let me have
it I shall be obliged.   When do you return ?—Ever
yours faithfully,   T. H. HUXLEY.

14 WAVERLEY PLACE, *Jan.* 1, 1859.

MY DEAREST LIZZIE—If intentions were only acts, the
quantity of letter-paper covered with my scrawl which
you would have had by this time would have been
something wonderful.   But I live at high pressure, with
always a number of things crying out to be done, and
those that are nearest and call loudest get done, while
the others, too often, don't.   However, this day shall not
go by without my wishing you all happiness in the new
year, and that wish you know necessarily includes all
belonging to you, and my love to them.

I have been long wanting to send you the photographs
of myself, wife, and boy, but one reason or other (Nettie's
incessant ill-health being, I am sorry to say, the chief)
has incessantly delayed the procuring of the last.   However, at length, we have obtained a tolerably successful
one, though you must not suppose that Noel has the
rather washed-out look of his portrait.   That comes of
his fair hair and blue gray eyes—for the monkey is like
his mother and has not an atom of resemblance to me.

He was two years old yesterday, and is the apple of
his father's eye and chief deity of his mother's pantheon,
which at present contains only a god and goddess.
Another is expected shortly, however, so that there is no
fear of Olympus looking empty.

. . . Here is the 26th of January and no letter gone
yet. . . . Since I began this letter I have been very
busy with lectures and other sorts of work, and besides,
my whole household almost has been ill—chicks with

whooping cough, mother with influenza, a servant ditto.
I don't know whether you have such things in Tennessee.

Let me see what has happened to me that will interest
you since I last wrote. Did I tell you that I have finally
made up my mind to stop in London—the Government
having made it worth my while to continue in Jermyn
Street? They give me £600 a year now, with a gradual
rise up to £800, which I reckon as just enough to live on
if one keeps very quiet. However, it is the greatest
possible blessing to be paid at last, and to be free from
all the abominable anxieties which attend a fluctuating
income. I can tell you I have had a sufficiently hard
fight of it.

When Nettie and I were young fools we agreed we
would marry whenever we had £200 a year. Well, we
have had more than twice that to begin upon, and how it
is we have kept out of the Bench is a mystery to me.
But we *have*, and I am inclined to think that the Missus
has got a private hoard (out of the puddings) for Noel.

I shall leave Nettie to finish this rambling letter. In
the meanwhile, my best love to you and yours, and mind
you are a better correspondent than your affectionate
brother,   Tom.

To Professor Leuckart

THE GOVERNMENT SCHOOL OF MINES,
JERMYN STREET, LONDON,
January, 30, 1859.

MY DEAR SIR—Our mutual friend, Dr. Harley, informs
me that you have expressed a wish to become possessed
of a separate copy of my lectures, published in the
Medical Times. I greatly regret that I have not one to
send you. The publisher only gave me half a dozen
separate copies of the numbers of the journal in which
the Lectures appeared. Of these I sent one to Johannes

Müller and one to Professor Victor Carus, and the rest went to other friends.

I am sorry to say that a mere fragment of what I originally intended to have published has appeared, the series having been concluded when I reached the end of the Crustacea. To say truth, the Lectures were not fitted for the journal in which they appeared.

I did not know that any one in Germany had noticed them until I received the copy of your *Bericht* for 1856, which you were kind enough to send me. I owe you many thanks for the manner in which you speak of them, and I assure you it was a source of great pleasure and encouragement to me to find so competent a judge as yourself appreciating and sympathising with my objects.

Particular branches of zoology have been cultivated in this country with great success, as you are well aware, but ten years ago I do not believe that there were half a dozen of my countrymen who had the slightest comprehension of morphology, and of what you and I should call "Wissenschaftliche Zoologie."

Those who thought about the matter at all took Owen's osteological extravaganzas for the *ne plus ultra* of morphological speculation.

I learned the meaning of Morphology and the value of development as the criterion of morphological views—first, from the study of the Hydrozoa during a long voyage, and secondly, from the writings of Von Bär. I have done my best, both by precept and practice, to inaugurate better methods and a better spirit than had long prevailed. Others have taken the same views, and I confidently hope that a new epoch for zoology is dawning among us. I do not claim for myself any great share in the good work, but I have not flinched when there was anything to be done.

Under these circumstances you will imagine that it was very pleasant to find on your side a recognition of what I was about.

I sent you, through the booksellers, some time ago, a copy of my memoir on Aphis.   I find from Moleschott's *Untersuchungen* that you must have been working at this subject contemporaneously with myself, and it was very satisfactory to find so close a concordance in essentials between our results.   Your memoirs are extremely interesting, and to some extent anticipated results at which my friend, Mr. Lubbock [1] (a very competent worker, with whose paper on Daphnia you are doubtless acquainted), had arrived.

I should be very glad to know what you think of my views of the composition of the articulate head.

I have been greatly interested also in your Memoir on Pentastomum.   There can be no difficulty about getting a notice of it in our journals, and, indeed, I will see to it myself.   Pray do me the favour to let me know whenever I can serve you in this or other ways.

I shall do myself the pleasure of forwarding to you immediately, through the booksellers, a lecture of mine on the Theory of the Vertebrate Skull, which is just published, and also a little paper on the development of the tail in fishes.

I am sorry to say that I have but little time for working at these matters now, as my position at the School of Mines obliges me to confine myself more and more to Paleontology.

However, I keep to the anatomical side of that sort of work, and so, now and then, I hope to emerge from amidst the fossils with a bit of recent anatomy.

Just at present, by the way, I am giving my disposable hours to the completion of a monograph on the Caly-cophoridæ and Physophoridæ observed during my voyage. The book ought to have been published eight years ago. But for three years I could get no money from the Government, and in the meanwhile you and Kölliker,

---

[1] The present Lord Avebury.

Gegenbaur and Vogt, went to the shores of the Mediterranean and made sad havoc with my novelties. Then came occupations consequent on my appointment to the chair I now hold ; and it was only last autumn that I had leisure to take up the subject again.

However, the plates, which I hope you will see in a few months, have, with two exceptions, been engraved five years.

Pray make my remembrances to Dr. Eckhard. I was sorry not to have seen him again in London.—Ever, my dear Sir, very faithfully yours,   T. H. HUXLEY.

Prof. Leuckart.

At this time Sir J. Hooker was writing, as an introduction to his *Flora of Tasmania,* his essay on the *Flora of Australia,* published in 1859—a book which owed its form to the influence of Darwin, and in return lent weighty support to evolutionary theory from the botanical side. He sent his proofs for Huxley to read.

14 WAVERLEY PLACE, N.W.,
April 22, 1859.

MY DEAR HOOKER—I have read your proofs with a great deal of attention and interest. I was greatly struck with the suggestions in the first page, and the exposure of the fallacy "that cultivated forms recur to wild types if left alone" is new to me and seems of vast importance.

The argument brought forward in the note is very striking and as simple as the egg of Columbus, when one sees it. I have marked one or two passages which are not quite clear to me. . . .

I have been accused of writing papers composed of nothing but heads of chapters, and I think you tend the same way. Please take the trouble to make the two lines I have scored into a paragraph, so that poor devils who are not quite so well up in the subject as yourself may not have to rack their brains for an hour to supply all the links of your chain of argument. . . .

You see that I am in a carping humour, but the matter of the essay seems to me to be so very valuable that I am jealous of the manner of it.

I had a long visit from Greene of Cork yesterday. He is very Irish, but very intelligent and well-informed, and I am in hopes he will do good service. He is writing a little book on the Protozoa, which (so far as I have glanced over the proof sheets as yet) seems to show a very philosophical turn of mind. It is very satisfactory to find the ideas one has been fighting for beginning to take root.

I do not suppose my own personal contributions to science will ever be anything very grand, but I shall be well content if I have reason to believe that I have done something to stir up others.—Ever yours faithfully,

T. H. HUXLEY.

To the same :—

April, 1859.

MY DEAR HOOKER . . . I pity you—as for the MSS. it is one of those cases for which penances were originally devised. What do you say to standing on your head in the garden for one hour per diem for the next week? It would be a relief. . . .

I suppose you will be at the Phil. Club next Monday. In the meanwhile don't let all the flesh be worried off your bones (there isn't much as it is).—Ever yours faithfully, T. H. HUXLEY.

14 WAVERLEY PLACE,
July 29, 1859.

MY DEAR HOOKER—I meant to have written to you yesterday, but things put it out of my head. If there is to be any fund raised at all, I am quite of your mind that it should be a scientific fund and not a mere naturalists' fund. Sectarianism in such matters is ridiculous, and besides that, in this particular case it is bad policy. For the word "Naturalist" unfortunately includes a far lower order of men than chemist, physicist, or mathematician. You don't call a man a mathematician because he has spent his life in getting as far as quadratics; but every fool who can make bad species and worse genera is a "Naturalist"!—save the mark! Imagine the chemists petitioning the Crown for a pension for P—— if he wanted one! and yet he really is a philosopher compared to poor dear A——.

"Naturalists" therefore are far more likely to want help than any other class of scientific men, and they would be greatly damaging their own interests if they formed an exclusive fund for themselves.—Ever yours faithfully,   T. H. HUXLEY.

CHAPTER XIII

1859

In November 1859 the *Origin of Species* was published, and a new direction was given to Huxley's activities. Ever since Darwin and Wallace had made their joint communication to the Linnean Society in July 1858, expectation had been rife as to the forthcoming book. Huxley was one of the few privileged to learn Darwin's argument before it was given to the world; but the greatness of the book, mere instalment as it was of the long accumulated mass of notes, almost took him by surprise. Before this time, he had taken up a thoroughly agnostic attitude with regard to the species question, for he could not accept the creational theory, yet sought in vain among the transmutationists for any cause adequate to produce transmutation. He had had many talks with Darwin, and though ready enough to accept the main point, maintained such a critical attitude on many others, that Darwin was not by any means certain of the effect the published book would produce upon him.

Indeed, in his 1857 notebook, I find jotted down
under the head of his paper on Pygocephalus (read
at the Geological Society), "anti-progressive confession of faith." Darwin was the more anxious, as,
when he first put pen to paper, he had fixed in his
mind three judges, by whose decision he determined
mentally to abide. These three were Lyell, Hooker,
and Huxley. If these three came round, partly
through the book, partly through their own reflections, he could feel that the subject was safe. "No
one," writes Darwin on November 13, "has read it,
except Lyell, with whom I have had much correspondence. Hooker thinks him a complete convert,
but he does not seem so in his letters to me; but
is evidently deeply interested in the subject." And
again : "I think I told you before that Hooker is a
complete convert. If I can convert Huxley I shall
be content." (*Life*, vol. ii. p. 221.)

On all three, the effect of the book itself, with its
detailed arguments and overwhelming array of evidence, was far greater than that of previous
discussions. With one or two reservations as to the
logical completeness of the theory, Huxley accepted
it as a well-founded working hypothesis, calculated
to explain problems otherwise inexplicable.

Two extracts from the chapter he contributed to
the *Life of Darwin* show very clearly his attitude
of mind when the *Origin of Species* was first
published :—

EXTRACT from "The Reception of the 'Origin of Species'" in *Life and Letters of Charles Darwin*, vol. ii. pp. 187-90 and 195-97.

I think I must have read the *Vestiges* before I left England in 1846 ; but, if I did, the book made very little impression upon me, and I was not brought into serious contact with the "Species" question until after 1850. At that time, I had long done with the Pentateuchal cosmogony, which had been impressed upon my childish understanding as Divine truth, with all the authority of parents and instructors, and from which it had cost me many a struggle to get free. But my mind was unbiassed in respect of any doctrine which presented itself, if it professed to be based on purely philosophical and scientific reasoning. It seemed to me then (as it does now) that "creation," in the ordinary sense of the word, is perfectly conceivable. I find no difficulty in conceiving that, at some former period, this universe was not in existence ; and that it made its appearance in six days (or instantaneously, if that is preferred), in consequence of the volition of some pre-existing Being. Then, as now, the so-called *a priori* arguments against Theism, and, given a Deity, against the possibility of creative acts, appeared to me to be devoid of reasonable foundation. I had not then, and I have not now, the smallest *a priori* objection to raise to the account of the creation of animals and plants given in *Paradise Lost*, in which Milton so vividly embodies the natural sense of Genesis. Far be it from me to say that it is untrue because it is impossible. I confine myself to what must be regarded as a modest and reasonable request for some particle of evidence that the existing species of animals and plants did originate in that way, as a condition of my belief in a statement which appears to me to be highly improbable.

And, by way of being perfectly fair, I had exactly the same answer to give to the evolutionists of 1851-58.

Within the ranks of the biologists, at that time, I met with nobody, except Dr. Grant of University College, who had a word to say for Evolution—and his advocacy was not calculated to advance the cause. Outside these ranks, the only person known to me whose knowledge and capacity compelled respect, and who was, at the same time, a thorough-going evolutionist, was Mr. Herbert Spencer, whose acquaintance I made, I think, in 1852, and then entered into the bonds of a friendship which, I am happy to think, has known no interruption. Many and prolonged were the battles we fought on this topic. But even my friend's rare dialectic skill and copiousness of apt illustration could not drive me from my agnostic position. I took my stand upon two grounds :—Firstly, that up to that time, the evidence in favour of trans-mutation was wholly insufficient ; and secondly, that no suggestion respecting the causes of transmutation assumed, which had been made, was in any way adequate to explain the phenomena. Looking back at the state of knowledge at that time, I really do not see that any other conclusion was justifiable.

In those days I had never even heard of Treviranus' *Biologie.* However, I had studied Lamarck attentively, and I had read the *Vestiges* with due care ; but neither of them afforded me any good ground for changing my negative and critical attitude. As for the *Vestiges,* I confess that the book simply irritated me by the prodigious ignorance and thoroughly unscientific habit of mind manifested by the writer. If it had any influence on me at all, it set me against Evolution ; and the only review I ever have qualms of conscience about, on the ground of needless savagery, is one I wrote on the *Vestiges* while under that influence. . . . [1]

---

[1] See the *British and Foreign Medico-Chirurgical Review,* xiii. 1854, p. 425, where the tenth edition of "this once attractive and still notorious work of fiction" is unsparingly examined, and its inaccurate statements and misty generalisations laid bare, until

But, by a curious irony of fate, the same influence
which led me to put as little faith in modern speculations
on this subject as in the venerable traditions recorded in
the first two chapters of Genesis, was perhaps more
potent than any other in keeping alive a sort of pious
conviction that Evolution, after all, would turn out true.
I have recently read afresh the first edition of the
Principles of Geology; and when I consider that this
remarkable book had been nearly thirty years in everybody's hands, and that it brings home to any reader of
ordinary intelligence a great principle and a great fact,—
the principle that the past must be explained by the
present, unless good cause be shown to the contrary ; and
the fact that so far as our knowledge of the past history
of life on our globe goes, no such cause can be shown,—
I cannot but believe that Lyell, for others, as for myself,
was the chief agent in smoothing the road for Darwin.
For consistent uniformitarianism postulates Evolution as
much in the organic as in the inorganic world.   The
origin of a new species by other than ordinary agencies

---

consistency and clearness are shown to belong as little to "our
Vestigiarian himself" as to "any of the other nebulæ."

The Review is further noteworthy (1) for its criticism, to be
reviewed more than thirty years later (see vol. iii. p. 11), in the cam-
paign against pseudo-science, of the "belief that a natural law is
an entity" instead of "nothing but the epitome of the observed
history of the phenomena of the universe " ; while on the other
hand "to assert that the Creator, from whom these phenomena
proceeded, worked in the manner of natural law, and that, therefore, there is no scope for wonder, is as if one should say that in
ancient Greece he worked in the manner of Grote's History, and
that, therefore, there is nothing remarkable in Greek civilisation."

It is noteworthy (2) for a strong attack on the "progressive"
theory of development, and shows that in face of geological and
biological evidence, it is untenable [cp. p. 247], and (3) for an ironical
rebuke to the author's open ascription to Owen of the authorship,
already attributed to him, of a *Quarterly* article—of course
unsigned—which not only ran counter to the views expressed in
his published works, but made an unjust attack on his nearest
colleague. (See p. 136, *supra*.)

would be a vastly greater "catastrophe" than any of those which Lyell successfully eliminated from sober geological speculation.

Thus, looking back into the past, it seems to me that my own position of critical expectancy was just and reasonable, and must have been taken up, on the same grounds, by many other persons. If Agassiz told me that the forms of life which have successively tenanted the globe were the incarnations of successive thoughts of the Deity, and that He had wiped out one set of these embodiments by an appalling geological catastrophe as soon as His ideas took a more advanced shape, I found myself not only unable to admit the accuracy of the deductions from the facts of paleontology, upon which this astounding hypothesis was founded, but I had to confess my want of any means of testing the correctness of his explanation of them. And besides that, I could by no means see what the explanation explained. Neither did it help me to be told by an eminent anatomist that species had succeeded one another in time, in virtue of "a continuously operative creational law." That seemed to me to be no more than saying that species had succeeded one another in the form of a vote-catching resolution, with "law" to catch the man of science, and "creational" to draw the orthodox. So I took refuge in that "thätige Skepsis" which Goethe has so well defined ; and, reversing the apostolic precept to be all things to all men, I usually defended the tenability of the received doctrines when I had to do with the transmutationists ; and stood up for the possibility of transmutation among the orthodox—thereby, no doubt, increasing an already current, but quite undeserved, reputation for needless combativeness.

I remember, in the course of my first interview with Mr. Darwin, expressing my belief in the sharpness of the lines of demarcation between natural groups and in the absence of transitional forms, with all the confidence of youth and imperfect knowledge. I was not aware, at that time, that he had then been many years brooding over the species-question; and the humorous smile which accompanied his gentle answer, that such was not altogether his view, long haunted and puzzled me. But it would seem that four or five years' hard work had enabled me to understand what it meant; for Lyell, writing to Sir Charles Bunbury (under date of April 30, 1856), says :—

"When Huxley, Hooker and Wollaston were at Darwin's last week, they (all four of them) ran a tilt against species—further, I believe, than they are prepared to go."

I recollect nothing of this beyond the fact of meeting Mr. Wollaston; and except for Sir Charles's distinct assurance as to "all four," I should have thought my *outrecuidance* was probably a counterblast to Wollaston's conservatism. With regard to Hooker, he was already, like Voltaire's Habbakuk, *capable de tout* in the way of advocating Evolution.

As I have already said, I imagine that most of those of my contemporaries who thought seriously about the matter were very much in my own state of mind— inclined to say to both Mosaists and Evolutionists, "a plague on both your houses!" and disposed to turn aside from an interminable and apparently fruitless discussion, to labour in the fertile fields of ascertainable fact. And I may therefore suppose that the publication of the Darwin and Wallace paper in 1858, and still more that of the "Origin" in 1859, had the effect upon them of the flash of light which, to a man who has lost himself on a dark night, suddenly reveals a road which, whether it takes him straight home or not, certainly goes his way. That which we were looking for, and could not find, was a hypothesis respecting the origin of known organic forms which assumed the operation of no causes but such as could be proved to be actually at work. We wanted, not to pin our faith to that or any other speculation, but to get hold of clear and definite conceptions which could be brought face to face with facts and have their validity tested. The "Origin" provided us with the working hypothesis we sought. Moreover, it did the immense service of freeing us for ever from the dilemma—Refuse to accept the creation hypothesis, and what have you to propose that can be accepted by any cautious reasoner? In 1857 I had no answer ready, and I do not think that any one else had. A year later we reproached ourselves with dulness for being perplexed with such an inquiry. My reflection, when I first made myself master of the central idea of the "Origin" was, "How extremely stupid not to have thought of that!" I suppose that Columbus' companions said much the same when he made the egg stand on end. The facts of variability, of the struggle for existence, of adaptation to conditions, were notorious enough ; but none of us had suspected that the road to the heart of the species problem lay through them, until Darwin and Wallace dispelled the darkness, and the beacon-fire of the "Origin" guided the benighted.

Whether the particular shape which the doctrine of Evolution, as applied to the organic world, took in Darwin's hands, would prove to be final or not, was to me a matter of indifference. In my earliest criticisms of the "Origin" I ventured to point out that its logical foundation was insecure so long as experiments in selective breeding had not produced varieties which were more or less infertile ; and that insecurity remains up to the present time. But, with any and every critical doubt which my sceptical ingenuity could suggest, the Darwinian hypothesis remained incomparably more probable than the creation hypothesis. And if we had none of us been able to discern the paramount significance of some of the most patent and notorious of natural facts, until they were, so to speak, thrust under our noses, what force remained in the dilemma—creation or nothing? It was obvious that hereafter the probability would be immensely greater, that the links of natural causation were hidden from our purblind eyes, than that natural causation should be incompetent to produce all the phenomena of nature. The only rational course for those who had no other object than the attainment of truth was to accept "Darwinism" as a working hypothesis and see what could be made of it. Either it would prove its capacity to elucidate the facts of organic life, or it would break down under the strain. This was surely the dictate of common sense, and, for once, common sense carried the day.

Even before the "Origin" actually came out, Huxley had begun to act as what Darwin afterwards called his "general agent." He began to prepare the way for the acceptance of the theory of evolution by discussing, for instance, one of the most obvious difficulties, namely, How is it that if evolution is ever progressive, progress is not universal? It was a point with respect to which Darwin himself wrote soon after the publication of the "Origin":— "Judging from letters . . . and from remarks, the most serious omission in my book was not explaining how it is, as I believe, that all forms do not necessarily advance, how there can now be *simple* organisms existing." (May 22, 1860.)

Huxley's idea, then, was to call attention to the persistence of many types without appreciable progression during geological time; to show that this fact was not explicable on any other hypothesis than that put forward by Darwin; and by paleontological arguments, to pave the way for consideration of the imperfection of the geological record.

Such were the lines on which he delivered his Friday evening lecture on "Persistent Types" at the Royal Institution on June 3, 1859.

However, the chief part which he took at this time in extending the doctrines of evolution was in applying them to his own subjects, Development and Vertebrate Anatomy, and more particularly to the question of the origin of mankind.

Of all the burning questions connected with the Origin of Species, this was the most heated—the most surrounded by prejudice and passion. To touch it was to court attack; to be exposed to endless scorn, ridicule, misrepresentation, abuse—almost to social ostracism. But the facts were there; the structural likenesses between the apes and man had already been shown; and as Huxley warned Darwin, "I will stop at no point so long as clear reasoning will carry me further."

Now two years before the "Origin" appeared, the denial of these facts by a leading anatomist led Huxley, as was his wont, to re-investigate the question for himself and satisfy himself one way or the other. He found that the previous investigators were not mistaken. Without going out of his way to refute the mis-statement as publicly as it was made, he simply embodied his results in his regular teaching. But the opportunity came unsought. Fortified by his own researches, he openly challenged these assertions when repeated at the Oxford meeting of the British Association in 1860, and promised to make good his challenge in the proper place.

We also find him combating some of the difficulties in the way of accepting the theory laid before him by Sir Charles Lyell. The veteran geologist had been Darwin's confidant from almost the beginning of his speculations; he had really paved the way for the evolutionary doctrine by his own proof of geological uniformity, but he shrank from accepting it, for its inevitable extension to the descent of man was repugnant to his feelings. Nevertheless, he would not allow sentiment to stand in the way of truth, and after the publication of the " Origin " it could be said of him—

Lyell, up to that time a pillar of the anti-transmutationists (who regarded him, ever after, as Pallas Athene may have looked at Dian, after the Endymion affair), declared himself a Darwinian, though not without putting in a serious *caveat*. Nevertheless, he was a tower of strength, and his courageous stand for truth as against consistency did him infinite honour.—(T. H. H. in *Life of Darwin*, vol. ii. p. 231.)

To Sir Charles Lyell

June 25, 1859

My dear Sir Charles—I have endeavoured to meet your objections in the enclosed.—Ever yours, very truly,

T. H. H.

The fixity and definite limitation of species, genera, and larger groups appear to me to be perfectly consistent with the theory of transmutation. In other words, I think *transmutation* may take place without transition.

Suppose that external conditions acting on species A give rise to a new species, B ; the difference between the two species is a certain definable amount which may be called A-B. Now I know of no evidence to show that the interval between the two species must *necessarily* be bridged over by a series of forms, each of which shall occupy, as it occurs, a fraction of the distance between A and B. On the contrary, in the history of the Ancon sheep, and of the six-fingered Maltese family, given by Reaumur, it appears that the new form appeared at once in full perfection.

I may illustrate what I mean by a chemical example. In an organic compound, having a precise and definite composition, you may effect all sorts of transmutations by substituting an atom of one element for an atom of another element. You may in this way produce a vast series of modifications—but each modification is definite in its composition, and there are no transitional or intermediate steps between one definite compound and another. I have a sort of notion that similar laws of definite combination rule over the modifications of organic bodies, and that in passing from species to species "Natura fecit saltum."

All my studies lead me to believe more and more in the absence of any real transitions between natural groups, great and small—but with what we know of the physiology of conditions [?] this opinion seems to me to be quite consistent with transmutation.

When I say that no evidence, or hardly any, would justify one in believing in the rise of a new species of Elephant, *e.g.* out of the earth, I mean that such an occurrence would be so diametrically contrary to all experience, so opposed to those beliefs which are the most constantly verified by experience, that one would be justified in believing either that one's senses were deluded, or that one had not really got to the bottom of the phenomenon. Of course, if one could vary the conditions, if one could take a little silex, and by a little *hocus-pocus* à la crosse, galvanise a baby out of it as often as one pleased, all the philosopher could do would be to hold up his hands and cry, "God is great." But short of evidence of this kind, I don't mean to believe anything of the kind.

How much evidence would you require to believe that there was a time when stones fell upwards, or granite made itself by a spontaneous rearrangement of the elementary particles of clay and sand? And yet the difficulties in the way of these beliefs are as nothing compared to those which you would have to overcome in believing that complex organic beings made themselves (for that is what creation comes to in scientific language) out of inorganic matter.

I know it will be said that even on the transmutation theory, the first organic being must have made itself. But there is as much difference between supposing the passage of inorganic matter into an *amoeba*, e.g., and into an *Elephant*, as there is between supposing that Portland stone might have built itself up into St. Paul's, and believing that the Giant's Causeway may have come about by natural causes.

True, one must believe in a beginning somewhere, but science consists in not believing the having reached that beginning before one is forced to do so.

It is wholly impossible to prove that any phenomenon whatsoever is not produced by the interposition of some unknown cause. But philosophy has prospered exactly as it has disregarded such possibilities, and has endeavoured to resolve every event by ordinary reasoning.

I do not exactly see the force of your argument that we are bound to find fossil forms intermediate between men and monkeys in the Rocks. Crocodiles are the highest reptiles as men are the highest mammals, but we find nothing intermediate between *crocodilia* and *lacertilia* in the whole range of the Mesozoic rocks. How do we know that Man is not a persistent type? And as for implements, at this day, and as, I suppose, for the last two or three thousand years at least, the savages of Australia have made their weapons of nothing but bone and wood. Why should *Homo Eocenus* or *Ooliticus*, the fellows who waddied the *Amphitherium* and speared the *Phascolotherium* as the Australian niggers treat their congeners, have been more advanced?

I by no means suppose that the transmutation hypothesis is proven or anything like it. But I view it as a powerful instrument of research. Follow it out, and it will lead us somewhere; while the other notion is like all the modifications of "final causation," a barren virgin.

And I would very strongly urge upon you that it is the logical development of Uniformitarianism, and that its adoption would harmonise the spirit of Paleontology with that of Physical Geology.

CHAPTER XIV

1859–1860

THE "Origin" appeared in November. As soon as he had read it, Huxley wrote the following letter to Darwin (already published in *Life of Darwin*, vol. ii. p. 231) :—

JERMYN STREET, W., *November* 23, 1859.

MY DEAR DARWIN—I finished your book yesterday, a lucky examination having furnished me with a few hours of continuous leisure.

Since I read Von Bar's essays, nine years ago, no work on Natural History Science I have met with has made so great an impression upon me, and I do most heartily thank you for the great store of new views you have given me. Nothing, I think, can be better than the tone of the book ; it impresses those who know about the subject. As for your doctrine, I am prepared to go to the stake, if requisite, in support of Chapter IX [1] and most parts of Chapters X, XI, XII, and Chapter XIII contains much that is most admirable, but on one or two

---

[1] Chapter IX, The Imperfection of the Geological Record ; X, The Geological Succession of Organic Beings ; XI-XII, Geographical Distribution ; XIII, Classification, Morphology, Embryology, and Rudimentary Organs.

points I enter a *caveat* until I can see further into all sides of the question.

As to the first four chapters,[1] I agree thoroughly and fully with all the principles laid down in them. I think you have demonstrated a true cause for the production of species, and have thrown the *onus probandi*, that species did not arise in the way you suppose, on your adversaries.

But I feel that I have not yet by any means fully realised the bearings of those most remarkable and original Chapters—III, IV, and V, and I will write no more about them just now.

The only objections that have occurred to me are—1st, That you have loaded yourself with an unnecessary difficulty in adopting *Natura non facit saltum* so unreservedly ; and 2nd, It is not clear to me why, if continual physical conditions are of so little moment as you suppose, variation should occur at all.

However, I must read the book two or three times more before I presume to begin picking holes.

I trust you will not allow yourself to be in any way disgusted or annoyed by the considerable abuse and misrepresentation which, unless I greatly mistake, is in store for you. Depend upon it, you have earned the lasting gratitude of all thoughtful men. And as to the curs which will bark and yelp, you must recollect that some of your friends, at anyrate, are endowed with an amount of combativeness which (though you have often and justly rebuked it) may stand you in good stead.

I am sharpening up my claws and beak in readiness.

Looking back over my letter, it really expresses so feebly all I think about you and your noble book, that I am half-ashamed of it ; but you will understand that, like the parrot in the story, "I think the more."—Ever yours faithfully,   T. H. HUXLEY.

---

[1] I, Variation under Domestication ; II, Variation under Nature ; III, The Struggle for Existence ; IV, Operation of Natural Selection ; V, Laws of Variation.

A month later, fortune put into his hands the opportunity of striking a vigorous and telling blow for the newly-published book. Never was windfall more eagerly accepted. A short account of this lucky chance was written by him for the Darwin *Life* (vol. i. p. 255).

The "Origin" was sent to Mr. Lucas, one of the staff of the *Times* writers at that day, in what was I suppose the ordinary course of business. Mr. Lucas, though an excellent journalist, and at a later period, editor of *Once a Week*, was as innocent of any knowledge of science as a babe, and bewailed himself to an acquaintance on having to deal with such a book. Whereupon, he was recommended to ask me to get him out of his difficulty, and he applied to me accordingly, explaining, however, that it would be necessary for him formally to adopt anything I might be disposed to write, by prefacing it with two or three paragraphs of his own.

I was too anxious to seize upon the opportunity thus offered of giving the book a fair chance with the multitudinous readers of the *Times*, to make any difficulty about conditions ; and being then very full of the subject, I wrote the article faster, I think, than I ever wrote anything in my life, and sent it to Mr. Lucas, who duly prefixed his opening sentences.

When the article appeared, there was much speculation as to its authorship. The secret leaked out in time, as all secrets will, but not by my aid ; and then I used to derive a good deal of innocent amusement from the vehement assertions of some of my more acute friends, that they knew it was mine from the first paragraph !

As the *Times*, some years since, referred to my connection with the review, I suppose there will be no breach of confidence in the publication of this little history, if you think it worth the space it will occupy.

The article appeared on December 26. Only Hooker was admitted into the secret. In an undated note Huxley writes to him :—

I have written the other review you wot of, and have handed it over to my friend to deal as he likes with it. . . . Darwin will laugh over a letter that I sent him this morning with a vignette of the Jermyn Street "pet" ready to fight his battle, and the "judicious Hooker" holding the bottle.

And on December 31 he writes again :—

JERMYN STREET, *December* 31, 1859.

MY DEAR HOOKER—I have not the least objection to my share in the *Times* article being known, only I should not like to have anything stated on my authority. The fact is, that the first quarter of the first column (down to "what is a species," etc.) is not mine, but belongs to the man who is the official reviewer for the *Times* (my "Temporal" godfather I might call him).

The rest is in my *ipsissima verba*, and I only wonder that it turns out as well as it does—for I wrote it faster than ever I wrote anything in my life. The last column nearly as fast as my wife could read the sheets. But I was thoroughly in the humour and full of the subject Of course as a scientific review the thing is worth nothing, but I earnestly hope it may have made some of the educated mob, who derive their ideas from the *Times*, reflect. And whatever they do, they *shall* respect Darwin.

Pray give my kindest regards and best wishes for the New Year to Mrs. Hooker, and tell her that if she, of her own natural sagacity and knowledge of the naughtiness of my heart, affirms that I wrote the article, I shall not contradict her—but that for reasons of state—I must not be supposed to say anything. I am pretty certain the Saturday article was not written by Owen. On internal grounds, because no word in it exceeds an inch in length ; on external, from what Cook said to me. The article is weak enough and one-sided enough, but looking at the various forces in action, I think Cook has fully redeemed his promise to me.

I went down to Sir P. Egerton on Tuesday—was ill when I started, got worse and had to come back on Thursday. I am all adrift now, but I couldn't stand being in the house any longer. I wish I had been born an *an-hepatous fœtus.*

All sorts of good wishes to you, and may you and I and Tyndalides, and one or two more bricks, be in as good fighting order in 1861 as in 1860.—Ever yours,

T. H. HUXLEY.

Speaking of this period and the half-dozen pre-ceding years, in his 1894 preface to *Man's Place in Nature* he says :—

Among the many problems which came under my consideration, the position of the human species in zoological classification was one of the most serious. Indeed, at that time it was a burning question in the sense that those who touched it were almost certain to burn their fingers severely. It was not so very long since my kind friend, Sir William Lawrence, one of the ablest men whom I have known, had been well-nigh ostracised for his book *On Man*, which now might be read in a Sunday school without surprising anybody ; it was only a few years since the electors to the chair of Natural History in a famous northern university had refused to invite a very distinguished man to occupy it because he advocated the doctrine of the diversity of species of mankind, or what was called "polygeny." Even among those who considered man from the point of view, not of vulgar prejudice, but of science, opinions lay poles asunder. Linnæus had taken one view, Cuvier another ; and among my senior contemporaries, men like Lyell, regarded by many as revolutionaries of the deepest dye, were strongly opposed to anything which tended to break down the barrier between man and the rest of the animal world.

My own mind was by no means definitely made up about this matter when, in the year 1857, a paper was read before the Linnæan Society " On the Characters, Principles of Division and Primary Groups of the Class Mammalia," in which certain anatomical features of the brain were said to be " peculiar to the genus ' Homo ' " and were made the chief ground for separating that genus from all other mammals and placing him in a division, " Archencephala," apart from, and superior to, all the rest. As these statements did not agree with the opinions I had formed, I set to work to reinvestigate the subject; and soon satisfied myself that the structures in question were not peculiar to Man, but were shared by him with all the higher and many of the lower apes. I embarked in no public discussion of these matters, but my attention being thus drawn to them, I studied the whole question of the structural relations of Man to the next lower existing forms with much care. And, of course, I embodied my conclusions in my teaching.

Matters were at this point when the *Origin of Species* appeared. The weighty sentence, " Light will be thrown on the origin of man and his history " (1st edition, p. 488), was not only in full harmony with the conclusions at which I had arrived respecting the structural relations of apes and men, but was strongly supported by them. And inasmuch as Development and Vertebrate Anatomy were not among Mr. Darwin's many specialities, it appeared to me that I should not be intruding on the ground he had made his own, if I discussed this part of the general question. In fact, I thought that I might probably serve the cause of Evolution by doing so.

Some experience of popular lecturing had convinced me that the necessity of making things clear to uninstructed people was one of the very best means of clearing up the obscure corners in one's own mind. So, in 1860, I took the Relation of Man to the Lower Animals for the subject of the six lectures to working men which it was my duty to deliver. It was also in 1860 that this topic was discussed before a jury of experts at the meeting of the British Association at Oxford, and from that time a sort of running fight on the same subject was carried on, until it culminated at the Cambridge Meeting of the Association in 1862, by my friend Sir W. Flower's public demonstration of the existence in the apes of those cerebral characters which had been said to be peculiar to man.

The famous Oxford Meeting of 1860 was of no small importance in Huxley's career. It was not merely that he helped to save a great cause from being stifled under misrepresentation and ridicule— that he helped to extort for it a fair hearing ; it was now that he first made himself known in popular estimation as a dangerous adversary in debate—a personal force in the world of science which could not be neglected. From this moment he entered the front fighting line in the most exposed quarter of the field.

Most unluckily, no contemporary account of his own exists of the encounter. Indeed, the same cause which prevented his writing home the story of the day's work nearly led to his absence from the scene. It was known that Bishop Wilberforce, whose first class in mathematics gave him, in popular estimation,

a right to treat on scientific matters, intended to
"smash Darwin"; and Huxley, expecting that the
promised debate would be merely an appeal to prejudice in a mixed audience, before which the
scientific arguments of the Bishop's opponents would
be at the utmost disadvantage, intended to leave Oxford
that very morning and join his wife at Hardwicke,
near Reading, where she was staying with her sister.
But in a letter, quoted below, he tells how, on the
Friday afternoon, he chanced to meet Robert
Chambers, the reputed author of the *Vestiges of
Creation*, who begged him "not to desert them."
Accordingly he postponed his departure; but seeing
his wife next morning, had no occasion to write a
letter.

Several accounts of the scene are already in
existence: one in the *Life of Darwin* (vol. ii. p. 320),
another in the 1892 *Life*, p. 236 *sq.*; a third that of
Lyell (vol. ii. p. 335), the slight differences between
them representing the difference between individual
recollections of eye-witnesses. In addition to these
I have been fortunate enough to secure further
reminiscences from several other eye-witnesses.

Two papers in Section D, of no great importance
in themselves, became historical as affording the
opponents of Darwin their opportunity of making an
attack upon his theory which should tell with the
public. The first was on Thursday, June 28. Dr.
Daubeny of Oxford made a communication to the
Section, "On the final causes of the sexuality of

plants, with particular reference to Mr. Darwin's work on the *Origin of Species*."[1] Huxley was called upon to speak by the President, but tried to avoid a discussion, on the ground "that a general audience, in which sentiment would unduly interfere with intellect, was not the public before which such a discussion should be carried on."

This consideration, however, did not stop the discussion; it was continued by Owen. He said he "wished to approach the subject in the spirit of the philosopher," and declared his "conviction that there were facts by which the public could come to some conclusion with regard to the probabilities of the truth of Mr. Darwin's theory." As one of these facts, he stated that the brain of the gorilla "presented more differences, as compared with the brain of man, than it did when compared with the brains of the very lowest and most problematical of the Quadrumana."

Now this was the very point, as said above, upon which Huxley had made special investigations during the last two years, with precisely opposite results, such as, indeed, had been arrived at by previous investigators. Hereupon he replied, giving these assertions a "direct and unqualified contradiction," and pledging himself to "justify that unusual procedure elsewhere," — a pledge which was amply

---

[1] My best thanks are due to Mr. F. Darwin for permission to quote his accounts of the meeting; other citations are from the *Athenæum* reports of July 14, 1860.

fulfilled in the pages of the *Natural History Review* for 1861.

Accordingly it was to him, thus marked out as the champion of the most debatable thesis of evolution, that, two days later, the Bishop addressed his sarcasms, only to meet with a withering retort. For on the Friday there was peace; but on the Saturday came a yet fiercer battle over the "Origin," which loomed all the larger in the public eye, because it was not merely the contradiction of one anatomist by another, but the open clash between Science and the Church. It was, moreover, not a contest of bare fact or abstract assertion, but a combat of wit between two individuals, spiced with the personal element which appeals to one of the strongest instincts of every large audience.

It was the merest chance, as I have already said, that Huxley attended the meeting of the section that morning. Dr. Draper of New York was to read a paper on the "Intellectual Development of Europe considered with reference to the views of Mr. Darwin." "I can still hear," writes one who was present, "the American accents of Dr. Draper's opening address when he asked 'Air we a fortuitous concourse of atoms?'" However, it was not to hear him, but the eloquence of the Bishop, that the members of the Association crowded in such numbers into the Lecture Room of the Museum, that this, the appointed meeting-place of the section, had to be abandoned for the long west room, since cut in two by a partition for the purposes of the library. It was not term time, nor were the general public admitted; nevertheless the room was crowded to suffocation long before the protagonists appeared on the scene, 700 persons or more managing to find places. The very windows by which the room was lighted down the length of its west side were packed with ladies, whose white handkerchiefs, waving and fluttering in the air at the end of the Bishop's speech, were an unforgettable factor in the acclamation of the crowd.

On the east side between the two doors was the platform. Professor Henslow, the President of the section, took his seat in the centre; upon his right was the Bishop, and beyond him again Dr. Draper; on his extreme left was Mr. Dingle, a clergyman from Lanchester, near Durham, with Sir J. Hooker and Sir J. Lubbock in front of him, and nearer the centre, Professor Beale of King's College, London, and Huxley.

The clergy, who shouted lustily for the Bishop, were massed in the middle of the room; behind them in the north-west corner a knot of undergraduates (one of these was T. H. Green, who listened but took no part in the cheering) had gathered together beside Professor Brodie, ready to lift their voices, poor minority though they were, for the opposite party. Close to them stood one of the few men among the audience already in Holy orders, who joined in—and indeed led—the cheers for the Darwinians.

So "Dr. Draper droned out his paper, turning first to the right hand and then to the left, of course bringing in a reference to the Origin of Species which set the ball rolling."

An hour or more that paper lasted, and then discussion began. The President "wisely announced *in limine* that none who had not valid arguments to bring forward on one side or the other would be allowed to address the meeting; a caution that proved necessary, for no fewer than four combatants had their utterances burked by him, because of their indulgence in vague declamation."[1]

First spoke (writes Professor Farrar[2]) a layman from Brompton, who gave his name as being one of the Committee of the (newly-formed) Economic section of the Association.   He, in a stentorian voice, let off his theological venom.   Then jumped up Richard Greswell[3] with a thin voice, saying much the same, but speaking as a scholar ; but we did not merely want any theological discussion, so we shouted them down.   Then a Mr. Dingle got up and tried to show that Darwin would have done much better if he had taken him into consultation. He used the blackboard and began a mathematical demonstration on the question—"Let this point A be man, and let that point B be the mawnkey."   He got no further ; he was shouted down with cries of "mawnkey." None of these had spoken more than three minutes.   It was when these were shouted down that Henslow said he must demand that the discussion should rest on *scientific* grounds only.

---

[1] *Life of Darwin, l.c.*
[2] Canon of Durham.
[3] Rev. Richard Greswell, B.D., Tutor of Worcester College.

Then there were calls for the Bishop, but he rose and said he understood his friend Professor Beale had something to say first. Beale, who was an excellent histologist, spoke to the effect that the new theory ought to meet with fair discussion, but added, with great modesty, that he himself had not sufficient knowledge to discuss the subject adequately. Then the Bishop spoke the speech that you know, and the question about his mother being an ape, or his grandmother.

From the scientific point of view, the speech was of small value. It was evident from his mode of handling the subject that he had been "crammed up to the throat," and knew nothing at first hand; he used no argument beyond those to be found in his *Quarterly* article, which appeared a few days later, and is now admitted to have been inspired by Owen. "He ridiculed Darwin badly and Huxley savagely; but," confesses one of his strongest opponents, "all in such dulcet tones, so persuasive a manner, and in such well-turned periods, that I who had been inclined to blame the President for allowing a discussion that could serve no scientific purpose, now forgave him from the bottom of my heart." [1]

The Bishop spoke thus "for full half an hour with inimitable spirit, emptiness and unfairness." "In a light, scoffing tone, florid and fluent, he assured us there was nothing in the idea of evolution; rock-pigeons were what rock-pigeons had always been. Then, turning to his antagonist with a smiling insolence, he begged to know, was it through his

[1] *Life of Darwin, l.c.*

grandfather or his grandmother that he claimed his descent from a monkey ?" [1]

This was the fatal mistake of his speech. Huxley instantly grasped the tactical advantage which the descent to personalities gave him. He turned to Sir Benjamin Brodie, who was sitting beside him, and emphatically striking his hand upon his knee, exclaimed, "The Lord hath delivered him into mine hands." The bearing of the exclamation did not dawn upon Sir Benjamin until after Huxley had completed his "forcible and eloquent" answer to the scientific part of the Bishop's argument, and proceeded to make his famous retort.[2]

[1] "Reminiscences of a Grandmother," *Macmillan's Magazine*, October 1898. Professor Farrar thinks this version of what the Bishop said is slightly inaccurate. His impression is that the words actually used seemed at the moment flippant and unscientific rather than insolent, vulgar, or personal. The Bishop, he writes, "had been talking of the perpetuity of species of Birds ; and then, denying *a fortiori* the derivation of the species Man from Ape, he rhetorically invoked the aid of *feeling*, and said, 'If any one were to be willing to trace his descent through an ape as his *grandfather*, would he be willing to trace his descent similarly on the side of his *grandmother*?' His false humour was an attempt to arouse the antipathy about degrading *woman* to the quadrumana. Your father's reply showed there was vulgarity as well as folly in the Bishop's words ; and the impression distinctly was, that the Bishop's party, as they left the room, felt abashed, and recognised that the Bishop had forgotten to behave like a perfect gentleman."

[2] The *Athenæum* reports him as saying that Darwin's theory was an explanation of phenomena in Natural History, as the undulatory theory was of the phenomena of light. No one objected to that theory because an undulation of light had never been arrested and measured. Darwin's theory was an explanation of facts, and his book was full of new facts, all bearing on his theory. Without asserting that every part of that theory had been confirmed, he maintained that it was the best explanation of the origin of species which had yet been offered. With regard to the psycho-

On this (continues the writer in *Macmillan's Magazine*)
Mr. Huxley slowly and deliberately arose. A slight tall
figure, stern and pale, very quiet and very grave,[1] he
stood before us and spoke those tremendous words—words
which no one seems sure of now, nor, I think, could
remember just after they were spoken, for their meaning
took away our breath, though it left us in no doubt as to
what it was. He was not ashamed to have a monkey for
his ancestor ; but he would be ashamed to be connected
with a man who used great gifts to obscure the truth.
No one doubted his meaning, and the effect was tremendous. One lady fainted and had to be carried out ; I, for
one, jumped out of my seat.

The fullest and probably most accurate account of
these concluding words is the following, from a letter
of the late John Richard Green, then an under-
graduate, to his friend, afterwards Professor Boyd
Dawkins [2] :—

logical distinction between men and animals, man himself was once
a monad—a mere atom, and nobody could say at what moment in
the history of his development he became consciously intelligent.
The question was not so much one of a transmutation or transition
of species, as of the production of forms which became permanent.
    Thus the short-legged sheep of America was not produced
gradually, but originated in the birth of an original parent of the
whole stock, which had been kept up by a rigid system of artificial
selection.
    [1] "Young, cool, quiet, scientific—scientific in fact and in treat-
ment."—J. R. Green. A certain piquancy must have been added
to the situation by the superficial resemblance in feature between
the two men, so different in temperament and expression. Indeed
next day at Hardwicke, a friend came up to Mr. Fanning and
asked who his guest was, saying, "Surely it is the son of the
Bishop of Oxford."
    [2] The writer in *Macmillan's* tells me : "I cannot quite accept
Mr. J. R. Green's sentences as your father's, though I didn't doubt
that they convey the sense ; but then I think that only a shorthand
writer could reproduce Mr. Huxley's singularly beautiful style—so
simple and so incisive. The sentence given is much too 'Green.'"

I asserted—and I repeat—that a man has no reason to be ashamed of having an ape for his grandfather. If there were an ancestor whom I should feel shame in recalling it would rather be a *man*—a man of restless and versatile intellect—who, not content with an equivocal [1] success in his own sphere of activity, plunges into scientific questions with which he has no real acquaintance, only to obscure them by an aimless rhetoric, and distract the attention of his hearers from the real point at issue by eloquent digressions and skilled appeals to religious prejudice. [2]

Further, Mr. A. G. Vernon - Harcourt, F.R.S., Reader in Chemistry at the University of Oxford, writes to me :—

The Bishop had rallied your father as to the descent from a monkey, asking as a sort of joke how recent this had been, whether it was his grandfather or further back. Your father, in replying on this point, first explained that the suggestion was of descent through thousands of generations from a common ancestor, and then went on to this effect—" But if this question is treated, not as a matter for the calm investigation of science, but as a matter of sentiment, and if I am asked whether I would choose to be descended from the poor animal of low intelligence and stooping gait, who grins and chatters as we pass, or from a man, endowed with great ability and a splendid position, who should use these gifts " [here, as the point became clear, there was a great outburst of applause,

---

[1] My father once told me that he did not remember using the word "equivocal" in this speech. (See his letter below.) The late Professor Victor Carus had the same impression, which is corroborated by Professor Farrar.

[2] As the late Henry Fawcett wrote in *Macmillan's Magazine*, 1860 :—" The retort was so justly deserved, and so inimitable in its manner, that no one who was present can ever forget the impression that it made."

which mostly drowned the end of the sentence] "to discredit and crush humble seekers after truth, I hesitate what answer to make."

No doubt your father's words were better than these, and they gained effect from his clear, deliberate utterance, but in outline and in *scale* this represents truly what was said.

After the commotion was over, "some voices called for Hooker, and his name having been handed up, the President invited him to give his view of the theory from the Botanical side. This he did, demonstrating that the Bishop, by his own showing, had never grasped the principles of the 'Origin,' and that he was absolutely ignorant of the elements of botanical science. The Bishop made no reply, and the meeting broke up." [1]

ACCOUNT OF THE OXFORD MEETING by the REV. W. H. FREMANTLE (in *Charles Darwin, his Life Told,* &c., 1892, p. 238).

The Bishop of Oxford attacked Darwin, at first playfully, but at last in grim earnest. It was known that the Bishop had written an article against Darwin in the last *Quarterly Review;* [2] it was also rumoured that Professor Owen had been staying at Cuddesdon and had primed the Bishop, who was to act as mouthpiece to the great Palæontologist, who did not himself dare to enter the lists. The Bishop, however, did not show himself master of the facts, and made one serious blunder. A fact which had been much dwelt on as confirmatory of Darwin's idea of varia-

---

[1] *Life of Darwin, l.c.*
[2] It appeared in the ensuing number for July.

tion, was that a sheep had been born shortly before in a flock in the North of England, having an addition of one to the vertebræ of the spine. The Bishop was declaring with rhetorical exaggeration that there was hardly any evidence on Darwin's side. "What have they to bring forward?" he exclaimed. "Some rumoured statement about a long-legged sheep." But he passed on to banter: "I should like to ask Professor Huxley, who is sitting by me, and is about to tear me to pieces when I have sat down, as to his belief in being descended from an ape. Is it on his grandfather's or his grandmother's side that the ape ancestry comes in?" And then taking a graver tone, he asserted, in a solemn peroration, that Darwin's views were contrary to the revelation of God in the Scriptures. Professor Huxley was unwilling to respond: but he was called for, and spoke with his usual incisiveness and with some scorn: "I am here only in the interests of science," he said, "and I have not heard anything which can prejudice the case of my august client." Then after showing how little competent the Bishop was to enter upon the discussion, he touched on the question of Creation. "You say that development drives out the Creator; but you assert that God made you: and yet you know that you yourself were originally a little piece of matter, no bigger than the end of this gold pencil-case." Lastly as to the descent from a monkey, he said: "I should feel it no shame to have risen from such an origin; but I should feel it a shame to have sprung from one who prostituted the gifts of culture and eloquence to the service of prejudice and of falsehood."

Many others spoke. Mr. Gresley, an old Oxford don, pointed out that in human nature at least orderly development was not the necessary rule: Homer was the greatest of poets, but he lived 3000 years ago, and has not produced his like.

Admiral FitzRoy was present, and said he had often expostulated with his old comrade of the *Beagle* for enter-

taining views which were contradictory to the First
Chapter of Genesis.

Sir John Lubbock declared that many of the arguments
by which the permanence of species was supported came
to nothing, and instanced some wheat which was said to
have come off an Egyptian mummy, and was sent to him
to prove that wheat had not changed since the time of the
Pharaohs ; but which proved to be made of French choco-
late. Sir Joseph (then Dr.) Hooker, spoke shortly, saying
that he had found the hypothesis of Natural Selection so
helpful in explaining the phenomena of his own subject
of Botany, that he had been constrained to accept it.
After a few words from Darwin's old friend, Professor
Henslow, who occupied the chair, the meeting broke
up, leaving the impression that those most capable of
estimating the arguments of Darwin in detail saw their
way to accept his conclusions.

Note.—Sir John Lubbock also insisted on the embryological evidence for evolution.　　　　　　　　F. D.

T. H. HUXLEY TO FRANCIS DARWIN (*ibid.*)

June 27, 1891.

I should say that Fremantle's account is substantially
correct, but that Green has the substance of my speech
more accurately. However, I am certain I did not use
the word " equivocal."

The odd part of the business is, that I should not
have been present except for Robert Chambers. I had
heard of the Bishop's intention to utilise the occasion.
I knew he had the reputation of being a first-class
controversialist, and I was quite aware that if he played
his cards properly, we should have little chance, with
such an audience, of making an efficient defence. Moreover, I was very tired, and wanted to join my wife at
her brother-in-law's country house near Reading, on the

Saturday. On the Friday I met Chambers in the street, and in reply to some remark of his, about his going to the meeting, I said that I did not mean to attend it—did not see the good of giving up peace and quietness to be episcopally pounded. Chambers broke out into vehement remonstrances, and talked about my deserting them. So I said, " Oh ! if you are going to take it that way, I'll come and have my share of what is going on."

So I came, and chanced to sit near old Sir Benjamin Brodie. The Bishop began his speech, and to my astonishment very soon showed that he was so ignorant that he did not know how to manage his own case. My spirits rose proportionately, and when he turned to me with his insolent question, I said to Sir Benjamin, in an undertone, " The Lord hath delivered him into mine hands."

That sagacious old gentleman stared at me as if I had lost my senses. But, in fact, the Bishop had justified the severest retort I could devise, and I made up my mind to let him have it. I was careful, however, not to rise to reply until the meeting called for me—then I let myself go.

In justice to the Bishop, I am bound to say he bore no malice, but was always courtesy itself when we occasionally met in after years. Hooker and I walked away from the meeting together, and I remember saying to him that this experience had changed my opinion as to the practical value of the art of public speaking, and that from that time forth I should carefully cultivate it, and try to leave off hating it. I did the former, but never quite succeeded in the latter effort.

I did not mean to trouble you with such a long scrawl when I began about this piece of ancient history.—Ever yours very faithfully, T. H. HUXLEY.

In the evening there was a crowded conversazione in Dr. Daubeny's rooms, and here, continues the writer in *Macmillan's*, " every one was eager to congratulate the hero of the day. I remember that some naïve person wished ' it could come over again ' ; Mr. Huxley, with the look on his face of the victor who feels the cost of victory, put us aside saying, ' Once in a lifetime is enough, if not too much.' "

In a letter to me the same writer remarks—

I gathered from Mr. Huxley's look when I spoke to him at Dr. Daubeny's that he was not quite satisfied to have been forced to take so personal a tone—it a little jarred upon his fine taste. But it was the Bishop who first struck the insolent note of personal attack.

Again, with reference to the state of feeling at the meeting :—

I never saw such a display of fierce party spirit, the looks of bitter hatred which the audience bestowed (I mean the majority) on us who were on your father's side ; as we passed through the crowd we felt that we were expected to say " how abominably the Bishop was treated " —or to be considered outcasts and detestable.

It was very different, however, at Dr. Daubeny's, " where," says the writer of the account in *Darwin's Life*, " the almost sole topic was the battle of the ' Origin,' and I was much struck with the fair and unprejudiced way in which the black coats and white cravats of Oxford discussed the question, and the frankness with which they offered their congratulations to the winners in the combat."

The result of this encounter, though a check to the other side, cannot, of course, be represented as an immediate and complete triumph for evolutionary doctrine. This was precluded by the character and temper of the audience, most of whom were less capable of being convinced by the arguments than shocked by the boldness of the retort, although, being gentlefolk, as Professor Farrar remarks, they were disposed to admit on reflection that the Bishop had erred on the score of taste and good manners. Nevertheless, it was a noticeable feature of the occasion, Sir M. Foster tells me, that when Huxley rose he was received coldly, just a cheer of encouragement from his friends, the audience as a whole not joining in it. But as he made his points the applause grew and widened, until, when he sat down, the cheering was not very much less than that given to the Bishop. To that extent he carried an unwilling audience with him by the force of his speech. The debate on the ape question, however, was continued elsewhere during the next two years, and the evidence was completed by the unanswerable demonstrations of Sir W. H. Flower at the Cambridge meeting of the Association in 1862.

The importance of the Oxford meeting lay in the open resistance that was made to authority, at a moment when even a drawn battle was hardly less effectual tnan acknowledged victory. Instead of being crushed under ridicule, the new theories secured a hearing, all the wider, indeed, for the startling nature of their defence.

CHAPTER XV

1860–1863

In the autumn he set to work to make good his promise of demonstrating the existence in the simian brain of the structures alleged to be exclusively human, and in particular of the *hippocampus minor*, a small eminence, shaped more or less like the sea-creature of that name, in the backward prolongation of the central hollow of the brain technically termed the posterior cornu of the lateral ventricle. The result was seen in his papers "On the Zoological Relations of Man with the Lower Animals" (*Nat. Hist. Rev.* 1861, pp. 67, 68); "On the Brain of Ateles Paniscus," which appeared in the *Proceedings of the Zoological Society* for 1861, and on "Nyctipithecus" in 1862; while similar work was undertaken by his friends Rolleston and Flower. But the brain was only one point among many, as, for example, the hand and the foot in man and the apes; and he already had in mind the discussion of the whole question comprehensively. On January 6 he writes to Sir J. Hooker :—

Some of these days I shall look up the ape question again and go over the rest of the organisation in the same way. But in order to get a thorough grip of the question I must examine into a good many points for myself. The results, when they do come out, will, I foresee, astonish the natives.

Full of interest in this theme, he made it the subject of his popular lectures in the spring of 1861.

Thus from February to May he lectured weekly to working men on "The Relation of Man to the rest of the Animal Kingdom," and on March 22 writes to his wife :—

My working men stick by me wonderfully, the house being fuller than ever last night. By next Friday evening they will all be convinced that they are monkeys. . . . Said lecture, let me inform you, was very good. Lyell came and was rather astonished at the magnitude and attentiveness of the audience.

These lectures to working men were published in the *Natural History Review*, as was a Friday evening discourse at the Royal Institution (February 8) on "The Nature of the Earliest Stages of Development of Animals."

Meanwhile the publication of these researches led to another pitched battle, in which public interest was profoundly engaged. The controversy which raged had some resemblance to a duel over a point of honour and credit. Scientific technicalities became the catchwords of society, and the echoes of the great Hippocampus question linger in the delightful pages

of the *Water-Babies*. Of this fight Huxley writes to
Sir J. Hooker on April 18, 1861 :—

A controversy between Owen and myself, which I can
only call absurd (as there is no doubt whatever about the
facts), has been going on in the *Athenæum*, and I wound
it up in disgust last week.

And again on April 27 :—

Owen occupied an entirely untenable position—but I
am nevertheless surprised he did not try "abusing
plaintiff's attorney." The fact is he made a prodigious
blunder in commencing the attack, and now his only
chance is to be silent and let people forget the exposure.
I do not believe that in the whole history of science there
is a case of any man of reputation getting himself into
such a contemptible position. He will be the laughing-
stock of all the continental anatomists.

Rolleston has a great deal of Oxford slough to shed,
but on that very ground his testimony has been of most
especial service. Fancy that man——telling Maskelyne
that Rolleston's observations were entirely confirmatory
of Owen.

About the same time he writes to his wife :—

April 16.—People are talking a good deal about the
"Man and the Apes" question, and I hear that somebody,
I suspect Monckton Milnes, has set afloat a poetical squib
on the subject.[1] . . . Some think my winding-up too
strong, but I trust the day will never come when I shall
abstain from expressing my contempt for those who
prostitute Science to the Service of Error. At any rate I

---

[1] The squib in question, dated "the Zoological Gardens," and
signed "Gorilla," appeared in *Punch* for May 15, 1861, under a
picture of that animal, bearing the sign, "*Am I a Man and a
Brother ?*"
The concluding verses run as follows :—

am not old enough for that yet. Darwin came in just now. I get no scoldings for pitching into the common enemy now !!

I would give you fifty guesses (he writes to Hooker on April 30), and you should not find out the author of the *Punch* poem. I saw it in MS. three weeks ago, and was told the author was a friend of mine. But I remained hopelessly in the dark till yesterday. What do you say to Sir Philip Egerton coming out in that line ? I am told he is the author, and the fact speaks volumes for Owen's perfect success in damning himself.

In the midst of the fight came a surprising invitation. On April 10 he writes to his wife : —

They have written to me from the Philosophical Institute of Edinburgh to ask me to give two lectures on the "Relation of Man to the Lower Animals" next session. I have replied that if they can give me January 3 and 7 for lecture days I will do it—if not, not. Fancy unco guid Edinburgh requiring illumination on the subject ! They know my views, so if they do not like what I shall have to tell them, it is their own fault.

These lectures were eventually delivered on January 4 and 7, 1862, and were well reported in the Edinburgh papers. The substance of them

---

Next *HUXLEY* replies
That *OWEN* he lies
And garbles his Latin quotation ;
That his facts are not new,
His mistakes not a few,
Detrimental to his reputation.

"To twice slay the slain"
By dint of the Brain
(Thus *HUXLEY* concludes his review)
Is but labour in vain,
Unproductive of gain,
And so I shall bid you "Adieu !

appears as Part 2 in *Man's Place in Nature*, the first
lecture describing the general nature of the process
of development among vertebrate animals, and the
modifications of the skeleton in the mammalia ; the
second dealing with the crucial points of comparison
between the higher apes and man, viz. the hand,
foot, and brain. He showed that the differences
between man and the higher apes were no greater
than those between the higher and lower apes. If
the Darwinian hypothesis explained the common
ancestry of the latter, the anatomist would have no
difficulty with the origin of man, so far as regards
the gap between him and the higher apes.

Yet, though convinced that "that hypothesis is as
near an approximation to the truth as, for example,
the Copernican hypothesis was to the true theory of
the planetary motions," he steadfastly refused to be
an advocate of the theory, "if by an advocate
is meant one whose business it is to smooth over
real difficulties, and to persuade when he cannot
convince."

In common fairness he warned his audience of the
one missing link in the chain of evidence—the fact
that selective breeding has not yet produced species
sterile to one another. But it is to be adopted as a
working hypothesis like other scientific generalisations, "subject to the production of proof that
physiological species may be produced by selective
breeding ; just as a physical philosopher may accept
the undulatory theory of light, subject to the proof

of the existence of the hypothetical ether ; or as the chemist adopts the atomic theory, subject to the proof of the existence of atoms ; and for exactly the same reasons, namely, that it has an immense amount of *prima facie* probability ; that it is the only means at present within reach of reducing the chaos of observed facts to order ; and lastly, that it is the most powerful instrument of investigation which has been presented to naturalists since the invention of the natural system of classification, and the commencement of the systematic study of embryology."

As for the repugnance of most men to admitting kinship with the apes, "thoughtful men," he says, "once escaped from the blinding influences of traditional prejudices, will find in the lowly stock whence man has sprung the best evidence of the splendour of his capacities ; and will discern, in his long progress through the past, a reasonable ground of faith in his attainment of a nobler future."

A simile used by him to enforce this elevating point of view, which has since eased the passage of many minds to the acceptance of evolution, seems to have been much appreciated by his audience. It was a comparison of man to the Alps, which turn out to be "of one substance with the dullest clay, but raised by inward forces to that place of proud and seemingly inaccessible glory."

The lectures were met at first with astonishing quiet but it was not long before the stones began to fly. The *Witness* of January 11 lashed itself into a

fury over the fact that the audience applauded this
"anti-scriptural and most debasing theory . . .
standing in blasphemous contradiction to biblical
narrative and doctrine," instead of expressing their
resentment at this "foul outrage committed upon
them individually, and upon the whole species as
'made in the likeness of God,'" by deserting the
hall in a body, or using some more emphatic form
of protest against the corruption of youth by "the
vilest and beastliest paradox ever vented in ancient
or modern times amongst Pagans or Christians." In
his finest vein of sarcasm, the writer expresses his
surprise that the meeting did not instantly resolve
itself into a "Gorilla Emancipation Society," or propose to hear a lecture from an apostle of Mormonism ;
"even this would be a less offensive, mischievous,
and inexcusable exhibition than was made in the
recent two lectures by Professor Huxley," etc.

JERMYN STREET, *January* 13, 1862.

MY DEAR DARWIN—In the first place a new year's
greeting to you and yours. In the next, I enclose this
slip (please return it when you have read it) to show you
what I have been doing in the north.

Everybody prophesied I should be stoned and cast out
of the city gate, but, on the contrary, I met with unmiti-
gated applause ! ! Three cheers for the progress of
liberal opinion ! !

The report is as good as any, but they have not put
quite rightly what I said about your views, respecting
which I took my old line about the infertility difficulty.

Furthermore, they have not reported my statement
that whether you were right or wrong, some form of the

progressive development theory is certainly true. Nor have they reported here my distinct statement that I believe man and the apes to have come from one stock.

Having got thus far, I find the lecture better reported in the *Courant*, so I send you that instead.

I mean to publish the lecture in full by and by (about the time the orchids come out [1]).—Ever yours faithfully,

T H HUXLEY.

I deserved the greatest credit for not having made an onslaught on Brewster for his foolish impertinence about your views in *Good Words*, but declined to stir nationality, which you know (in him) is rather more than his Bible.

JERMYN STREET, *January* 16, 1862.

MY DEAR HOOKER—I wonder if we are ever to meet again in this world ! At any rate I send to the remote province of Kew, Greeting, and my best wishes for the new year to you and yours. I also enclose a slip from an Edinburgh paper containing a report of my lecture on the "Relation of Man," etc. As you will see, I went in for the entire animal more strongly, in fact, than they have reported me. I told them in so many words that I entertained no doubt of the origin of man from the same stock as the apes.

And to my great delight, in saintly Edinburgh itself the announcement met with nothing but applause. For myself I can't say that the praise or blame of my audience was much matter, but it is a grand indication of the general disintegration of old prejudices which is going on.

I shall see if I cannot make something more of the lectures by delivering them again in London, and then I shall publish them.

The report does not put nearly strong enough what I said in favour of Darwin's views. I affirmed it to be the only scientific hypothesis of the origin of species

---

[1] *i.e.* Darwin's book on Orchids.

in existence, and expressed my belief that the one gap
in the evidence would be filled up, as I always do.—
Ever yours faithfully,   T. H. HUXLEY.

JERMYN STREET, *January* 20, 1862.

MY DEAR DARWIN—The enclosed article, which has been
followed up by another more violent, more scurrilously
personal, and more foolish, will prove to you that my
labour has not been in vain, and that your views and
mine are likely to be better ventilated in Scotland than
they have been.

I was quite uneasy at getting no attack from the
Witness, thinking I must have overestimated the impression I had made, and the favourableness of the reception
of what I said. But the raving of the *Witness* is clear
testimony that my notion was correct.

I shall send a short reply to the *Scotsman* for the
purpose of further advertising the question.

With regard to what are especially your doctrines, I
spoke much more favourably than I am reported to have
done. I expressed no doubt as to their ultimate establishment, but as I particularly wished not to be misrepresented
as an advocate trying to soften or explain away real
difficulties, I did not in speaking enter into the details
of what is to be said in diminishing the weight of the
hybrid difficulty. All this will be put fully when I
print the Lecture.

The arguments put in your letter are those which I
have urged to other people—of the opposite side—over
and over again. I have told my students that I entertain
no doubt that twenty years' experiments on pigeons
conducted by a skilled physiologist, instead of by a
mere breeder, would give us physiological species sterile
inter se, from a common stock (and in this, if I mistake
not, I go further than you do yourself), and I have
told them that when these experiments have been
performed I shall consider your views to have a complete

physical basis, and to stand on as firm ground as any physiological theory whatever.

It was impossible for me, in the time I had, to lay all this down to my Edinburgh audience, and in default of full explanation it was far better to seem to do scanty justice to you. I am constitutionally slow of adopting any theory that I must needs stick by when I have once gone in for it; but for these two years I have been gravitating towards your doctrines, and since the publication of your primula paper with accelerated velocity. By about this time next year I expect to have shot past you, and to find you pitching into me for being more Darwinian than yourself. However, you have set me going, and must just take the consequences, for I warn you I will stop at no point so long as clear reasoning will carry me further.

My wife and I were very grieved to hear you had had such a sick house, but I hope the change in the weather has done you all good. Anything is better than the damp warmth we had.

I will take great care of the three "Barriers."[1] I wanted to cut it up in the *Saturday*, but how I am to fulfil my benevolent intentions—with five lectures a week—a lecture at the Royal Institution and heaps of other things on my hands, I don't know.—Ever yours faithfully,   T. H. HUXLEY.

I am very glad to hear about Brown Sequard; he is a thoroughly good man, and told me it was worth while to come all the way to Oxford to hear the Bishop pummelled.

In the above-mentioned letter to the *Scotsman* of January 24 he expresses his unfeigned satisfaction

[1] A pamphlet called "The Three Barriers, by G. R., being notes on Mr. Darwin's *Origin of Species*, 1861, 8vo." Habitat, structure, and procreative power are given as these three barriers to Darwinism, against which natural theology takes its stand on Final Causes.

at the fulfilment of the three objects of his address, namely, to state fully and fairly his conclusions, to avoid giving unnecessary offence, and thirdly, "while feeling assured of the just and reasonable dealing of the respectable part of the Scottish press, I naturally hoped for noisy injustice and unreason from the rest, seeing, as I did, the best security for the dissemination of my views through regions which they might not otherwise reach, in the certainty of a violent attack by (the *Witness*)."

The applause of the audience, he says, afforded him genuine satisfaction, "because it bids me continue in the faith on which I acted, that a man who speaks out honestly and fearlessly that which he knows, and that which he believes, will always enlist the good-will and the respect, however much he may fail in winning the assent, of his fellow-men."

About this time a new field of interest was opened out to him, closely connected with, indeed, and completing, the ape question. Sir Charles Lyell was engaged in writing his *Antiquity of Man*, and asked Huxley to supply him with various anatomical data touching the ape question, and later to draw him a diagram illustrating the peculiarities of the newly-discovered Neanderthal skull as compared with other skulls. He points out in his letters to Lyell that the range of cranial capacity between the highest and the lowest German—"one of the mediatised princes, I suppose"[1]—or the Malayan or Peruvian, is almost

[1] The minor princes of Germany, whose territories were annexed

100 per cent; in absolute amount twice as much as the difference between that of the largest simian and the smallest human capacity, so that in seeking an ordinal difference between man and the apes, "it would certainly be well to let go the head, though I am afraid it does not mend matters much to lay hold of the foot."

And on January 25, 1862 :—

I have been skull-measuring all day at the College of Surgeons. The *Neanderthal skull* may be described as a slightly exaggerated modification of one of the two types (and the lower) of Australian skulls.

After the fashion of accounting for the elephant of old, I suppose it will be said that it was imported. But luckily the differences, though only of degree, are rather too marked for this hypothesis.

I only wish I had a clear six months to work at the subject. Little did I dream what the undertaking to arrange your three woodcuts would lead to. It will come in the long-run, I believe, to a new ethnological method, new modes of measurement, a new datum line, and new methods of registration.

If one had but two heads and neither required sleep!

One immediate result of his investigations, which appeared in a lecture at the Royal Institution (February 7, 1862), "On the Fossil Remains of Man," was incorporated in *Man's Place in Nature.* But a more important consequence of this impulse was that he went seriously into the study of Ethnology. Of his work in this branch of natural

to larger states, and who thus exchanged a direct for a mediate share in the imperial government.

science, Professor Virchow, speaking at the dinner given him by the English medical profession on October 5, 1898, declared that in the eyes of German savants it alone would suffice to secure immortal reverence for his name.

The concluding stage in the long controversy raised first at Oxford was the British Association meeting at Cambridge in 1862. It was here that Professor (afterwards Sir W. H.) Flower made his public demonstration of the existence in apes of the cerebral characters said to be peculiar to man.

From the 1st to the 9th of October Huxley stayed at Cambridge as the guest of Professor Fawcett at Trinity Hall, running over to Felixstow on the 5th to see his wife, whose health did not allow her to accompany him.

As President of Section D he had a good deal to do, and he describes the course of events in a letter to Darwin :—

26 ABBEY PLACE, *Oct.* 9, 1862.

MY DEAR DARWIN—It is a source of sincere pleasure to me to learn that anything I can say or do is a pleasure to you, and I was therefore very glad to get your letter at that whirligig of an association meeting the other day. We all missed you, but I think it was as well you did not come, for though I am pretty tough, as you know, I found the pace rather killing. Nothing could exceed the hospitality and kindness of the University people—and that, together with a great deal of speaking on the top of a very bad cold, which I contrived to catch just before going down, has somewhat used me up.

Owen came down with the obvious intention of attacking me on all points. Each of his papers was an attack, and he went so far as to offer stupid and unnecessary opposition to proposals of mine in my own committee. However, he got himself sold at all points. . . . The Polypterus paper and the Aye-Aye paper fell flat. The latter was meant to raise a discussion on your views, but it was all a stale hash, and I only made some half sarcastic remarks which stopped any further attempts at discussion. . . .

I took my book to Scotland but did nothing. I shall ask leave to send you a bit or two as I get on.—Ever yours,   T. H. HUXLEY.

A "Society for the propagation of common honesty in all parts of the world" was established at Cambridge. I want you to belong to it, but I will say more about it by and by.

This admirable society, which was also to "search for scientific truth, especially in biology," seems to have been but short lived. At all events, I can find only two references to subsequent meetings, on October 7 and December 19 in this year.

A few days later a final blow was struck in the battle over the ape question. He writes on October 15 how he has written a letter to the *Medical Times*—his last word on the subject, summing up in most emphatic terms :—

I have written the letter with the greatest care, and there is nothing coarse or violent in it. But it shall put an end to all the humbug that has been going on. . . . Rolleston will come out with his letter in the same number, and the smash will be awful, but most thoroughly merited.

These several pieces of work, struck out at different times in response to various impulses, were now combined and re-shaped into *Man's Place in Nature*, the first book which was published by him. Thus he writes to Sir Charles Lyell on May 5, 1862 :—

Of course I shall be delighted to discuss anything with you,[1] and the more so as I mean to put the whole question before the world in another shape in my little book, whose title is announced as *Evidences as to Man's Place in Nature.* I have written the two first essays, the second containing the substance of my Edinburgh Lecture. I recollect you once asked me for something to quote on the Man question, so if you want anything in that way the MS. is at your service.

Lyell looked over the proofs, and the following letters are in reply to his criticisms :—

ARDRISHAIG, LOCH FYNE,
Aug. 17, 1862.

MY DEAR SIR CHARLES—I take advantage of my first quiet day to reply to your letter of the 9th ; and in the first place let me thank you very much for your critical remarks, as I shall find them of great service.

With regard to such matters as verbal mistakes, you must recollect that the greater part of the proof was wholly uncorrected. But the reader might certainly do his work better. I do not think you will find room to complain of any want of distinctness in my definition of Owen's position touching the Hippocampus question. I mean to give the whole history of the business in a note, so that the paraphrase of Sir Ph. Egerton's line " To

---

[1] Referring to the address on "Geological Contemporaneity" delivered in 1862 at the Geological Society, see p. 296.

which Huxley replies that Owen he lies," shall be un-mistakable.[1]

I will take care about the Cheiroptera, and I will look at Lamarck again. But I doubt if I shall improve my estimate of the latter. The notion of common descent was not his—still less that of modification by variation—and he was as far as De Maillet from seeing his way to any *vera causa* by which varieties might be intensified into species.

If Darwin is right about natural selection—the dis-covery of this *vera causa* sets him to my mind in a different region altogether from all his predecessors—and I should no more call his doctrine a modification of Lamarck's than I should call the Newtonian theory of the celestial motions a modification of the Ptolemaic system. Ptolemy imagined a mode of explaining those motions. Newton proved their necessity from the laws and a force demonstrably in operation. If he is only right Darwin will, I think, take his place with such men as Harvey, and even if he is wrong his sobriety and accuracy of thought will put him on a far different level from Lamarck. I want to make this clear to people.

I am disposed to agree with you about the "emasculate" and "uncircumcised"—partly for your reasons, partly because I believe it is an excellent rule always to erase anything that strikes one as particularly smart when writing it. But it is a great piece of self-denial to abstain from expressing my peculiar antipathy to the people indicated, and I hope I shall be rewarded for the virtue.

As to the secondary causes I only wished to guard myself from being understood to imply that I had any comprehension of the meaning of the term. If my phrase looks naughty I will alter it. What I want is to be read, and therefore to give no unnecessary handle to the enemy. There will be row enough whatever I do.

[1] See p. 278.

Our Commission here [1] implicates us in an inquiry of some difficulty, and which involves the interests of a great many poor people. I am afraid it will not leave me very much leisure. But we are in the midst of a charming country, and the work is not unpleasant or uninteresting. If the sun would only shine more than once a week it would be perfect.—With kind remembrances to Lady Lyell, believe me, faithfully yours,

T. H. HUXLEY.

We shall be here for the next ten days at least. But my wife will always know my whereabouts.

JERMYN STREET, *March* 23, 1863.

MY DEAR SIR CHARLES—I suspect that the passage to which you refer must have been taken from my unrevised proofs, for it corresponds very nearly with what is written at p. 97 of my book.

Flower has recently discovered that the Siamang's brain affords an even more curious exception to the general rule than that of *Mycetes*, as the cerebral hemispheres leave part not only of the sides but of the hinder end of the cerebellum uncovered.

As it is one of the Anthropoid apes and yet differs in this respect far more widely from the gorilla than the gorilla differs from man, it offers a charming example of the value of cerebral characters.

Flower publishes a paper on the subject in the forthcoming number of the *N. H. Review*.

Might it not be well to allude to the fact that the existence of the posterior lobe, posterior cornu, and hippocampus in the Orang has been publicly demonstrated to an audience of experts at the College of Surgeons ?—Ever yours faithfully,   T. H. HUXLEY.

The success of *Man's Place* was immediate, despite

---

[1] The Fishery Commission.

such criticisms as that of the *Athenæum,* that "Lyell's object is to make man old, Huxley's to degrade him." By the middle of February it reached its second thousand ; in July it is heard of as republished in America ; at the same time L. Buchner writes that he wished to translate it into German, but finds himself forestalled by Victor Carus. From another aspect, Lord Enniskillen, thanking him for the book, says (March 3), " I believe you are already excommunicated by book, bell, and candle," while in an undated note, Bollaert writes, " The Bishop of Oxford the other day spoke about ' the church having been in danger of late, by such books as Colenso's, but that it (the church) was now restored.' And this at a time, he might have added, when the works of Darwin, Lyell, and Huxley are torn from the hands of Mudie's shop-men, as if they were novels—(see *Daily Telegraph,* April 10)."

At the same time, the impression left by his work upon the minds of the leading men of science may be judged from a few words of Sir Charles Lyell, who writes to a friend on March 15, 1863 (*Life and Letters,* ii. 366) :—

Huxley's second thousand is going off well. If he had leisure like you and me, and the vigour and logic of the lectures, and his address to the Geological Society, and half a dozen other recent works (letters to the *Times* on Darwin, etc.), had been all in one book, what a position he would occupy ! I entreated him not to undertake the *Natural History Review* before it began. The responsibility all falls on the man of chief energy and talent ; it is a quarterly mischief, and will end in knocking him up.

A similar estimate appears from an earlier letter of March 11, 1859 (*Life and Letters*, ii. 321), when he quotes Huxley's opinion of Mansel's Bampton Lectures on the *Limits of Religious Thought* :—

> A friend of mine, Huxley, who will soon take rank as one of the first naturalists we have ever produced, begged me to read these sermons as first rate, "although, regarding the author as a churchman, you will probably compare him, as I did, to the drunken fellow in Hogarth's contested election, who is sawing through the signpost at the other party's public-house, forgetting he is sitting at the other end of it. But read them as a piece of clear and unanswerable reasoning."

In the 1894 preface to the re-issue of *Man's Place* in the Collected Essays, Huxley speaks as follows of the warnings he received against publishing on so dangerous a topic, of the storm which broke upon his head, and the small result which, in the long run, it produced [1] :—

> *Magna est veritas et prœvalebit !* Truth is great, certainly, but considering her greatness, it is curious what a long time she is apt to take about prevailing. When, towards the end of 1862, I had finished writing *Man's Place in Nature*, I could say with a good conscience that my conclusions "had not been formed hastily or enunciated crudely." I thought I had earned the right to

---

[1] In September 1887 he wrote to Mr. Edward Clodd—"All the propositions laid down in the wicked book, which was so well anathematised a quarter of a century ago, are now taught in the text-books. What a droll world it is ! "

publish them, and even fancied I might be thanked
rather than reproved for doing so.   However, in my
anxiety to publish nothing erroneous, I asked a highly
competent anatomist and very good friend of mine to look
through my proofs, and, if he could, point out any errors
of fact.   I was well pleased when he returned them without criticism on that score ; but my satisfaction was
speedily dashed by the very earnest warning as to the
consequences of publication, which my friend's interest in
my welfare led him to give.   But, as I have confessed
elsewhere, when I was a young man, there was just a
little—a mere *soupçon*—in my composition of that tenacity
of purpose which has another name ; and I felt sure that
all the evil things prophesied would not be so painful to
me as the giving up that which I had resolved to do,
upon grounds which I conceived to be right.[1]   So the
book came out ; and I must do my friend the justice to
say that his forecast was completely justified.   The Boreas
of criticism blew his hardest blasts of misrepresentation
and ridicule for some years, and I was even as one of the
wicked.   Indeed, it surprises me at times to think how
any one who had sunk so low could since have emerged
into, at any rate, relative respectability.   Personally, like
the non-corvine personages in the Ingoldsby legend, I did
not feel " one penny the worse."   Translated into several
languages, the book reached a wider public than I had
ever hoped for ; being largely helped, I imagine, by the
Ernulphine advertisements to which I referred.   It has
had the honour of being freely utilised without acknow-

---

[1] As to this advice not to publish *Man's Place* for fear of misrepresentation on the score of morals, he said, in criticising an
attack of this sort made upon Darwin in the *Quarterly* for July
1876 :—" It seemed to me, however, that a man of science has no
raison d'être at all, unless he is willing to face much greater risks
than these for the sake of that which he believes to be true ; and
further, that to a man of science such risks do not count for much
—that they are by no means so serious as they are to a man of
letters, for example."

ledgment by writers of repute; and finally it achieved
the fate, which is the euthanasia of a scientific work, of
being enclosed among the rubble of the foundations of
later knowledge, and forgotten.

To my observation, human nature has not sensibly
changed during the last thirty years. I doubt not that
there are truths as plainly obvious and as generally denied
as those contained in *Man's Place in Nature,* now await-
ing enunciation. If there is a young man of the present
generation who has taken as much trouble as I did to
assure himself that they are truths, let him come out
with them, without troubling his head about the barking
of the dogs of St. Ernulphus. *Veritas prævalebit*—some
day; and even if she does not prevail in his time, he
himself will be all the better and wiser for having tried
to help her. And let him recollect that such great
reward is full payment for all his labour and pains.

The following letter refers to the newly-published
Man's Place in Nature. Miss H. Darwin had suggested
a couple of corrections :—

JERMYN STREET, *Feb.* 25, 1863.

MY DEAR DARWIN—Please to say to Miss Henrietta
Minos Rhadamanthus Darwin that I plead guilty to the
justice of both criticisms, and throw myself on the mercy
of the court.

As extenuating circumstances with respect to indict-
ment No. 1, see prefatory notice. Extenuating circumstance No. 2—that I picked up "Atavism" in Pritchard
years ago, and as it is a much more convenient word than
"Hereditary transmission of variations," it slipped into
equivalence in my mind, and I forgot all about the
original limitation.

But if these excuses should in your judgment tend to
aggravate my offences, suppress 'em like a friend. One

may always hope more from a lady's tender-heartedness than from her sense of justice.

Publisher has just sent to say that I must give him any corrections for second thousand of my booklet immediately.

Why did not Miss Etty send any critical remarks on that subject by the same post? I should be most immensely obliged for them.—Ever yours faithfully,

T. H. HUXLEY.

During this period of special work at the anthropological side of the Evolution theory, Huxley made two important contributions to the general question.

As secretary of the Geological Society, the duty of delivering the anniversary address in 1862 fell to him in the absence of the president, Leonard Horner, who had been driven by ill-health to winter in Italy.

The object at which he aimed appears from the postscript of a brief note of Feb. 19, 1862, to Hooker:—

I am writing the body of the address, and I am going to criticise Palæontological doctrines in general in a way that will flutter their nerves considerable.

Darwin is met everywhere with—Oh this is opposed to palæontology, or that is opposed to palæontology—and I mean to turn round and ask, "Now, Messieurs les Palæontologues, what the devil do you really know?"

I have not changed sex although the postscript is longer than the letter.

The delivery of the address [1] itself on February 21 is thus described by Sir Charles Lyell [2] (*Life and Letters*, ii. 356) :—

---

[1] On "Geological Contemporaneity" (*Coll. Ess.* viii. 292).

[2] To a note of whose, proposing a talk over the subject, Huxley replies on May 5, "I am very glad you find something to think about in my address. That is the best of all praise."

Huxley delivered a brilliant critical discourse on what palæontology has and has not done, and proved the value of negative evidence, how much the progressive development system has been pushed too far, how little can be said in favour of Owen's more generalised types when we go back to the vertebrata and invertebrata of remote ages, the persistency of many forms high and low throughout time, how little we know of the beginning of life upon the earth, how often events called contemporaneous in Geology are applied to things which, instead of coinciding in time, may have happened ten millions of years apart, etc. ; and a masterly sketch comparing the past and present in almost every class in zoology, and sometimes of botany cited from Hooker, which he said he had done because it was useful to look into the cellars and see how much gold there was there, and whether the quantity of bullion justified such an enormous circulation of paper. I never remember an address listened to with such applause, though there were many private protests against some of his bold opinions.

The dinner at Willis's was well attended ; I should think eighty or more present . . . and late in the evening Huxley made them merry by a sort of mock-modest speech.

JERMYN STREET, *May* 6, 1862.

MY DEAR DARWIN—I was very glad to get your note about my address. I profess to be a great stoic, you know, but there are some people from whom I am glad to get a pat on the back. Still I am not quite content with that, and I want to know what you think of the argument—whether you agree with what I say about contemporaneity or not, and whether you are prepared to admit—as I think your views compel you to do—that the whole Geological Record is only the skimmings of the pot of life.

Furthermore, I want you to chuckle with me over the

notion I find a great many people entertain—that the
address is dead against your views. The fact being, as
they will by and by wake up [to] see that yours is the
only hypothesis which is not negatived by the facts,—one
of its great merits being that it allows not only of indefinite standing still, but of indefinite retrogression.

I am going to try to work the whole argument into an
intelligible form for the general public as a chapter of my
forthcoming "Evidence"[1] (one-half of which I am happy
to say is now written), so I shall be very glad of any criticisms or hints.

Since I saw you—indeed, from the following Tuesday
onwards—I have amused myself by spending ten days
or so in bed. I had an unaccountable prostration of
strength which they called influenza, but which, I believe
was nothing but some obstruction in the liver.

Of course I can't persuade people of this, and they
will have it that it is overwork. I have come to the
conviction, however, that steady work hurts nobody, the
real destroyer of hard-working men being not their work,
but dinners, late hours, and the universal humbug and
excitement of society.

I mean to get out of all that and keep out of it.—Ever
yours faithfully,   T. H. HUXLEY.

The other contribution to the general question
was his Working Men's Lectures for 1862. As he
writes to Darwin on October 10—"I can't find anything to talk to the working men about this year but
your book. I mean to give them a commentary à la
Coke upon Lyttleton."

The lectures to working men here referred to, six
in number, were duly delivered once a week from
November 10 onwards, and published in the form of

[1] *Evidence as to Man's Place in Nature.*

as many little pamphlets. Appearing under the general title, "On our Knowledge of the Causes of the Phenomena of Organic Nature," they wound up with a critical examination of the portion of Mr. Darwin's work *On the Origin of Species*, in relation to the complete theory of the causes of organic nature.

JERMYN STREET, *Dec.* 2, 1862.

MY DEAR DARWIN—I send you by this post three of my working men's lectures now in course of delivery. As you will see by the prefatory notice, I was asked to allow them to be taken down in shorthand for the use of the audience, but I have no interest in them, and do not desire or intend that they should be widely circulated.

Some time hence, may be, I may revise and illustrate them, and make them into a book as a sort of popular exposition of your views, or at any rate of my version of your views.

There really is nothing new in them nor anything worth your attention, but if in glancing over them at any time you should see anything to object to, I should like to know.

I am very hard worked just now—six lectures a week, and no end of other things—but as vigorous as a three-year old. Somebody told me you had been ill, but I hope it was fiction, and that you and Mrs. Darwin and all your belongings are flourishing.—Ever yours faithfully,

T. H. HUXLEY.

In reply, Darwin writes on December 10:—

I agree entirely with all your reservations about accepting the doctrine, and you might have gone further with perfect safety and truth. . . .

Touching the *Natural History Review*, "Do inaugurate

a great improvement, and have pages cut, like the Yankees do ; I will heap blessings on your head."

And again, December 18 :—

I have read Nos. IV. and V.   They are simply perfect. They ought to be largely advertised ; but it is very good in me to say so, for I threw down No. IV. with this reflection, " What is the good of my writing a thundering big book, when everything is in this green little book so despicable for its size ? "   In the name of all that is good and bad I may as well shut up shop altogether.

These lectures met with an annoying amount of success.   They were not cast into permanent form, for he grudged the time necessary to prepare them for the press.   However, he gave a Mr. Hardwicke permission to print them from a shorthand report for the use of the audience.   But no sooner were they printed, than they had a large sale.   Writing to Sir J. D. Hooker early in the following month, he says :—

I fully meant to have sent you all the successive lectures as they came out, and I forward a set with all manner of apologies for my delinquency.   I am such a 'umble-minded party that I never imagined the lectures as delivered would be worth bringing out at all, and I knew I had no time to work them out.   Now, I lament I did not publish them myself and turn an honest penny by them as I suspect Hardwicke is doing.   He is advertising them everywhere, confound him.

I wish when you have read them you would tell me whether you think it would be worth while for me to re-edit. enlarge, and illustrate them by and by.

And on January 28 Sir C. Lyell writes to him :—

I do grudge Hardwicke very much having not only the publisher's but the author's profits. It so often happens that popular lectures designed for a class and inspired by an attentive audience's sympathy are better than any writing in the closet for the purpose of educating the many as readers, and of remunerating the publisher and author. I would lose no time in considering well what steps to take to rescue the copyright of the third thousand.

As for the value of the work thus done in support of Darwin's theory, it is worth while quoting the words of Lord Kelvin, when, as President of the Royal Society in 1894, it fell to him to award Huxley the Darwin Medal :—

To the world at large, perhaps, Mr. Huxley's share in moulding the thesis of *Natural Selection* is less well known than is his bold unwearied exposition and defence of it after it had been made public. And, indeed, a speculative trifler, revelling in the problems of the "might have been," would find a congenial theme in the inquiry how soon what we now call "Darwinism" would have met with the acceptance with which it has met, and gained the power which it has gained, had it not been for the brilliant advocacy with which in its early days it was expounded to all classes of men.

That advocacy had one striking mark : while it made or strove to make clear how deep the new view went down, and how far it reached, it never shrank from trying to make equally clear the limit beyond which it could not go.

CHAPTER XVI

1860-1861

THE letters given in the following chapters illustrate the occupations and interests of the years 1860 to 1863, apart from the struggle over the species question.

One of the most important and most engrossing was the launching of a scientific quarterly to do more systematically and thoroughly what had been done since 1858 in the fortnightly scientific column of the *Saturday Review*. Its genesis is explained in the following letter :—

July 17, 1860.

MY DEAR HOOKER—Some time ago Dr. Wright of Dublin talked to me about the *Natural History Review*, which I believe to a great extent belongs to him, and wanted me to join in the editorship, provided certain alterations were made. I promised to consider the matter, and yesterday he and Greene dined with me, and I learned that Haughton and Galbraith were out of the review—that Harvey was likely to go—that a new series was to begin in January, with Williams and Norgate for publishers over here—that it was to become

an English and not a Hibernian concern in fact—and
finally, that if I chose to join as one of the editors, the
effectual control would be pretty much in my own
hands. Now, considering the state of the times, and the
low condition of natural history journalisation (always
excepting quarterly *Mic. Jour.*) in this country, this
seems to me to be a fine opening for a plastically minded
young man, and I am decidedly inclined to close with
the offer, though I shall get nothing but extra work by it.

To limit the amount of this extra work, however, I
must get co-editors, and I have written to Lubbock and
to Rolleston (also plastically minded young men) to see
if they will join. Now up to this point you have been
in a horrid state of disgust, because you thought I was
going to ask you next. But I am not, for rejoiced as I
should be to have you, I know you have heaps of better
work to do, and hate journalism.

But can you tell me of any plastic young botanist
who would come in all for glory and no pay, though I
think pay may be got if the concern is properly worked.
How about Oliver ?

And though you can't and won't be an editor yourself,
won't you help us and pat us on the back.

The tone of the *Review* will be mildly episcopophagous,
and you and Darwin and Lyell will have a fine opportunity
if you wish it of slaying your adversaries.—Ever yours
faithfully,   T. H. HUXLEY.

Several of his elder friends tried to dissuade him
from an undertaking which would inevitably distract
him from his proper work. Sir C. Lyell prophesied
(see p. 294) that all the work would drift to the
most energetic member of the staff, and Huxley
writes to Hooker, August 2, 1860 :—

Darwin wrote me a very kind expostulation about it,
telling me I ought not to waste myself on other than

original work. In reply, however, I assured him that
I *must* waste myself willy-nilly, and that the *Review* was
only a save all.

The more I think of it the more it seems to me it
ought to answer if properly conducted, and it ought to
be of great use.

The first number appeared in January 1861.
Writing on the 6th, Huxley says :—

It is pleasant to get such expressions of opinion as I
have had from Lyell and Darwin about the *Review*. They
make me quite hopeful about its prosperity, as I am sure
we shall be able to do better than our first number.

It was not long, however, before Lyell's prophecy
began to come true. In June Huxley writes :—

It is no use letting other people look after the journal.
I find unless I revise every page of it, it goes wrong.

But in July 1863 he definitely ceased to contribute :—

I did not foresee all this crush of work (he writes)
when the *Review* was first started, or I should not have
pledged myself to any share in supplying it. (Moreover,
with the appointment of paid editors that year, it seemed
to him) that the working editors with the credit and the
pay must take the responsibility of all the commissariat
of the *Review* upon their shoulders.

Two years later, in 1865, the *Review* came to an
end. As Mr. Murray, the publisher, remarked,
quarterlies did not pay ; and this quarterly became
still more financially unsound after the over-worked

volunteers, who both edited and contributed, gave place to paid editors.

But Huxley was not satisfied with one defeat. The quarterly scheme had failed ; he now tried if he could not serve science better by returning to a more frequent and more popular form of periodical. From 1863 to 1866 he was concerned with the *Reader*, a weekly issue ;[1] but this also was too heavy a burden to be borne in addition to his other work. However, the labour expended in these ventures was not wholly thrown away. The experience thus gained at last enabled the present Sir Norman Lockyer, who acted as science editor for the *Reader*, to realise what had so long been aimed at by the establishment of *Nature* in 1869.

Apart from his contributions to the species question and the foundation of a scientific review, Huxley published in 1860 only two special monographs (" On Jacare and Caiman," and " On the Mouth and Pharynx of the Scorpion," already mentioned as read in the previous year), but he read "Further Observations on Pyrosoma" at the Linnean Society, and was busy with paleontological work, the results of which appeared in three papers the following year, the most important of which was the Memoir called a " Preliminary Essay on the Arrangement of the Devonian Fishes," in the report of the Geological Survey, "which," says Sir M. Foster,

[1] The committee also included Professor Cairns, F. Galton, W. F. Pollock, and J. Tyndall.

"though entitled a Preliminary Essay, threw an entirely new light on the affinities of these creatures, and, with the continuation published later, in 1866, still remains a standard work."

The question of the admission of ladies to the learned societies was already being mooted, and a letter to Sir C. Lyell gives his ideas thus early not only on this point, but on the general question of women's education.

March 17, 1860.

MY DEAR SIR CHARLES—To use the only forcible expression, I "twig" your meaning perfectly, but I venture to think the parable does not apply. For the Geological Society is not, to my mind, a place of education for students, but a place of discussion for adepts; and the more it is applied to the former purpose the less competent it must become to fulfil the latter—its primary and most important object.

I am far from wishing to place any obstacle in the way of the intellectual advancement and development of women. On the contrary, I don't see how we are to make any permanent advancement while one-half of the race is sunk, as nine-tenths of women are, in mere ignorant parsonese superstitions; and to show you that my ideas are practical, I have fully made up my mind, if I can carry out my own plans, to give my daughters the same training in physical science as their brother will get, so long as he is a boy. They, at any rate, shall not be got up as man-traps for the matrimonial market. If other people would do the like, the next generation would see women fit to be the companions of men in all their pursuits—though I don't think that men have anything to fear from their competition. But you know as well as I do that other people won't do the like, and five-sixths of women will stop in the doll stage of evolution to be the stronghold of parsondom, the drag on civilisation, the degradation of every important pursuit with which they mix themselves—"intrigues" in politics, and "friponnes" in science.

If my claws and beak are good for anything, they shall be kept from hindering the progress of any science I have to do with.—Ever yours faithfully,

T. H. HUXLEY.

Three letters to Mr. Spencer show that he had been reading and criticising biological points in the proofs of *First Principles*. With regard to the second letter, which gives reasons for rejecting Mr. Spencer's remarks about the power of inflation in birds during flight, it is curious to note Mr. Spencer's reply :—

How oddly the antagonism comes out even when you are not conscious of it! My authority was Owen! I heard him assign this cause for the falling of wounded birds in one of his lectures at the College of Surgeons.

14 WAVERLEY PLACE, *Sept.* 3, 1860.

MY DEAR SPENCER—I return your proofs by this post. To my mind nothing can be better than their contents, whether in matter or in manner, and as my wife arrived, independently, at the same opinion, I think my judgment is not one-sided.

There is something calm and dignified about the tone of the whole—which eminently befits a philosophical work which means to live—and nothing can be more clear and forcible than the argument.

I rejoice that you have made a beginning, and such a beginning—for the more I think about it the more important it seems to me that somebody should think out into a connected system the loose notions that are floating about more or less distinctly in all the best minds.

It seems as if all the thoughts in what you have written were my own, and yet I am conscious of the enormous difference your presentation of them makes in my intellectual state. One is thought in the state of hemp yarn, and the other in the state of rope. Work away, then, excellent rope-maker, and make us more ropes to hold on against the devil and the parsons.

For myself, I am absorbed in dogs—gone to the dogs in fact—having been occupied in dissecting them for the last fortnight. You do not say how your health is.—Ever yours faithfully,   T. H. HUXLEY.

Sept. 19, 1860.

MY DEAR SPENCER—You will forgive the delay which has occurred in forwarding your proof when I tell you that we have lost our poor little son, our pet and hope. You who knew him well, and know how his mother's heart and mine were wrapped up in him, will understand how great is our affliction. He was attacked with a bad form of scarlet fever on Thursday night, and on Saturday night effusion on the brain set in suddenly and carried him ill in a couple of hours. Jessie was taken ill on Friday, but has had the disease quite lightly, and is doing well. The baby has escaped. So end many hopes and plans—sadly enough, and yet not altogether bitterly. For as the little fellow was our greatest joy, so is the recollection of him an enduring consolation. It is a heavy payment, but I would buy the four years of him again at the same price. My wife bears up bravely.

I have read your proofs at intervals, and you must not suppose they have troubled me. On the contrary they were at times the only things I could attend to. I agree in the spirit of the whole perfectly. On some matters of detail I had doubts which I am not at present clear-headed enough to think out.

The only thing I object to *in toto* is the illustration which I have marked at p. 24. It is physically impossible that a bird's air-cells should be *distended* with air during flight, unless the structure of the parts is in reality different from anything which anatomists at present know. Blowing into the trachea is not to the point. A bird cannot blow into its own trachea, and it has no mechanism for performing a corresponding action.

A bird's chest is essentially a pair of bellows, in which the sternum during rest and the back during flight act as movable walls. The air-cells may all be represented as soft-walled bags opening freely into the bellows— there being, so far as anatomists yet know, no valves or corresponding contrivances anywhere except at the glottis, which corresponds with the nozzle and air-valve both, of our bellows. But the glottis is always opened when the chest is dilated at each inspiration. How then can the air in any air-cell be kept at a higher tension than the surrounding atmosphere?

Hunter experimented on the uses of the air-sacs, I know, but I have not his works at hand. It may be that opening one of the air-cells interferes with flight, but I hold it very difficult to conceive that the interference can take place in the way you suppose. How on earth is a lark to sing for ten minutes together if the air-cells are to be kept distended all the while he is up in the air?

At any rate twenty other illustrations will answer your purpose as well, so I would not select one which may be assailed by a carping fellow like—Yours very faithfully,

T. H. HUXLEY.

Oct. 10, 1860.

MY DEAR SPENCER [1]—"A wilful man must have his way," and if you won't let me contribute towards the material guarantees for the success of your book, I must be content to add twelve shillings' worth of moral

---

[1] Mr. Spencer had insisted that he should be on the free list, instead of being a paying subscriber to the serial issue of the *Synthetic Philosophy.*

influence to that I already meant to exert per annum in its favour.

I shall be most glad henceforth, as ever, to help your great undertaking in any way I can. The more I contemplate its issues the more important does it seem to me to be, and I assure you that I look upon its success as the business of all of us. So that if it were not a pleasure, I should feel it a duty to "push behind" as hard as I can.

Have you seen this quarter's *Westminster*? The opening article on "Neo-Christianity" is one of the most remarkable essays in its way I have ever read. I suppose it must be Newman's. The *Review* is terribly unequal, some of the other articles being absolutely ungrammatically written. What a pity it is it cannot be thoroughly organised.

My wife is a little better, but she is terribly shattered. By the time you come back we shall, I hope, have reverted from our present hospital condition to our normal arrangements, but in any case we shall be glad to see you.— Ever yours faithfully,   T. H. HUXLEY.

The following is, I think, the first reference to his fastidiousness in the literary expression and artistic completeness of his work. As he said in an after-dinner speech at a meeting in aid of the Literary Fund, "Science and literature are not two things, but two sides of one thing." Anything that was to be published he subjected to repeated revision. And thus, apologising to Hooker for his absence, he writes (August 2, 1860)—

I was sorry to have to send an excuse by Tyndall the other day, but I found I must finish the Pyrosoma paper, and all last Tuesday was devoted to it, and I fear the next after will have the like fate.

It constantly becomes more and more difficult to me to *finish* things satisfactorily.

To Hooker also he writes a few days later :—

I hope your ear is better ; take care of yourself, there's a good fellow. I can't do without you these twenty years. We have a devil of a lot to do in the way of smiting the Amalekites.

Between two men who seldom spoke of their feelings, but let constant intercourse attest them, these words show more than the practical side of their friendship, their community of aims and interests. Quick, strong-willed, and determined as they both were, the fact that they could work together for over forty years without the shadow of a misunderstanding, presupposes an unusually strong friendship firmly based upon mutual trust and respect as well as liking, the beginning of which Sir J. Hooker thus describes :—

My first meeting your father was in 1851, shortly after his return from the *Rattlesnake* voyage with Captain Stanley. Hearing that I had paid some attention to marine zoology during the voyage of the Antarctic Expedition, he was desirous of showing me the results of his studies of the Oceanic Hydrozoa, and he sought me out in consequence. This and the fact that we had both embarked in the Naval service in the same capacity as medical officers and with the same object of scientific research, naturally led to an intimacy which was undisturbed by a shadow of a misunderstanding for nearly forty-five following years. Curiously enough, our intercourse might have dated from an earlier period by nearly six years had I accepted an appointment to the *Rattle-snake* offered me by Captain Stanley, which, but for my having arranged for a journey to India, might have been accepted. Returning to the purpose of our interview, the researches Mr. Huxley laid before me were chiefly those on the Salpæ, a much misunderstood group of marine Hydrozoa. Of these I had amused myself with making drawings during the long and often weary months passed at sea on board the *Erebus*, but having other subjects to attend to, I had made no further study of them than as consumers of the vegetable life (Diatoms) of the Antarctic Ocean. Hence his observations on their life-history, habits, and affinities were on almost all points a revelation to me, and I could not fail to recognise in their author all the qualities possessed by a naturalist of commanding ability, industry, and power of exposition. Our interviews, thus commenced, soon ripened into a friendship, which led to an arrangement for a monthly meeting, and in the informal establishment of a club of nine, the other members of which were, Mr. Busk, Dr. Frankland, Mr. Hirst, Sir J. Lubbock, Mr. Herbert Spencer, Dr. Tyndall, and Mr. Spottiswoode.

Just a month after this letter to his friend, the same year which had first brought Huxley public recognition outside his special sphere brought him also the greatest sorrow perhaps of his whole life. I have already spoken of the sudden death of the little son in whom so much of his own and his wife's happiness was centred. The suddenness of the blow made it all the more crushing, and the mental strain, intensified by the sight of his wife's inconsolable grief, brought him perilously near a complete break-down. But the birth of another son, on December

11, gave the mother some comfort; and as the result of a friendly conspiracy between her and Dr. Tyndall, Huxley himself was carried off for a week's climbing in Wales between Christmas and the New Year.

His reply to a long letter of sympathy in which Charles Kingsley set forth the grounds of his own philosophy as to the ends of life and the hope of immortality, affords insight into the very depths of his nature. It is a rare outburst at a moment of intense feeling, in which, more completely than in almost any other writing of his, intellectual clearness and moral fire are to be seen uniting in a veritable passion for truth :—

14 WAVERLEY PLACE, Sept. 23, 1860.

MY DEAR KINGSLEY—I cannot sufficiently thank you, both on my wife's account and my own, for your long and frank letter, and for all the hearty sympathy which it exhibits—and Mrs. Kingsley will, I hope, believe that we are no less sensible of her kind thought of us. To myself your letter was especially valuable, as it touched upon what I thought even more than upon what I said in my letter to you. My convictions, positive and negative, on all the matters of which you speak, are of long and slow growth and are firmly rooted. But the great blow which fell upon me seemed to stir them to their foundation, and had I lived a couple of centuries earlier I could have fancied a devil scoffing at me and them—and asking me what profit it was to have stripped myself of the hopes and consolations of the mass of mankind ? To which my only reply was and is—Oh devil ! truth is better than much profit. I have searched over the grounds of my belief, and if wife and child and name and fame were all to be lost to me one after the other as the penalty, still I will not lie.

And now I feel that it is due to you to speak as frankly as you have done to me. An old and worthy friend of mine tried some three or four years ago to bring us together—because, as he said, you were the only man who would do me any good. Your letter leads me to think he was right, though not perhaps in the sense he attached to his own words.

To begin with the great doctrine you discuss. I neither deny nor affirm the immortality of man. I see no reason for believing in it, but, on the other hand, I have no means of disproving it.

Pray understand that I have no *a priori* objections to the doctrine. No man who has to deal daily and hourly with nature can trouble himself about *a priori* difficulties. Give me such evidence as would justify me in believing anything else, and I will believe that. Why should I not? It is not half so wonderful as the conservation of force, or the indestructibility of matter. Whoso clearly appreciates all that is implied in the falling of a stone can have no difficulty about any doctrine simply on account of its marvellousness. But the longer I live, the more obvious it is to me that the most sacred act of a man's life is to say and to feel, "I believe such and such to be true." All the greatest rewards and all the heaviest penalties of existence cling about that act. The universe is one and the same throughout; and if the condition of my success in unravelling some little difficulty of anatomy or physiology is that I shall rigorously refuse to put faith in that which does not rest on sufficient evidence, I cannot believe that the great mysteries of existence will be laid open to me on other terms. It is no use to talk to me of analogies and probabilities. I know what I mean when I say I believe in the law of the inverse squares, and I will not rest my life and my hopes upon weaker convictions. I dare not if I would.

Measured by this standard, what becomes of the doctrine of immortality ?

You rest in your strong conviction of your personal existence, and in the instinct of the persistence of that existence which is so strong in you as in most men.

To me this is as nothing. That my personality is the surest thing I know—may be true. But the attempt to conceive what it is leads me into mere verbal subtleties. I have champed up all that chaff about the ego and the non-ego, about noumena and phenomena, and all the rest of it, too often not to know that in attempting even to think of these questions, the human intellect flounders at once out of its depth.

It must be twenty years since, a boy, I read Hamilton's essay on the unconditioned, and from that time to this, ontological speculation has been a folly to me. When Mansel took up Hamilton's argument on the side of orthodoxy (!) I said he reminded me of nothing so much as the man who is sawing off the sign on which he is sitting, in Hogarth's picture. But this by the way.

I cannot conceive of my personality as a thing apart from the phenomena of my life. When I try to form such a conception I discover that, as Coleridge would have said, I only hypostatise a word, and it alters nothing if, with Fichte, I suppose the universe to be nothing but a manifestation of my personality. I am neither more nor less eternal than I was before.

Nor does the infinite difference between myself and the animals alter the case. I do not know whether the animals persist after they disappear or not. I do not even know whether the infinite difference between us and them may not be compensated by *their* persistence and *my* cessation after apparent death, just as the humble bulb of an annual lives, while the glorious flowers it has put forth die away.

Surely it must be plain that an ingenious man could speculate without end on both sides, and find analogies for all his dreams. Nor does it help me to tell me that the aspirations of mankind—that my own highest aspirations even — lead me towards the doctrine of immortality. I doubt the fact, to begin with, but if it be so even, what is this but in grand words asking me to believe a thing because I like it.

Science has taught to me the opposite lesson. She warns me to be careful how I adopt a view which jumps with my preconceptions, and to require stronger evidence for such belief than for one to which I was previously hostile.

My business is to teach my aspirations to conform themselves to fact, not to try and make facts harmonise with my aspirations.

Science seems to me to teach in the highest and strongest manner the great truth which is embodied in the Christian conception of entire surrender to the will of God. Sit down before fact as a little child, be prepared to give up every preconceived notion, follow humbly wherever and to whatever abysses nature leads, or you shall learn nothing   l have only begun to learn content and peace of mind since I have resolved at all risks to do this.

There are, however, other arguments commonly brought forward in favour of the immortality of man, which are to my mind not only delusive but mischievous. The one is the notion that the moral government of the world is imperfect without a system of future rewards and punishments. The other is : that such a system is indispensable to practical morality. I believe that both these dogmas are very mischievous lies.

With respect to the first, I am no optimist, but I have the firmest belief that the Divine Government (if we may use such a phrase to express the sum of the " customs of matter ") is wholly just. The more I know intimately of the lives of other men (to say nothing of my own), the more obvious it is to me that the wicked does *not* flourish ncr is the righteous punished. But for this to be clear we must bear in mind what almost all forget, that the rewards of life are contingent upon obedience to the *whole* law—physical as well as moral—and that moral obedience will not atone for physical sin, or *vice versa*.

The ledger of the Almighty is strictly kept, and every one of us has the balance of his operations paid over to him at the end of every minute of his existence.

Life cannot exist without a certain conformity to the surrounding universe—that conformity involves a certain amount of happiness in excess of pain. In short, as we live we are paid for living.

And it is to be recollected in view of the apparent discrepancy between men's acts and their rewards that Nature is juster than we. She takes into account what a man brings with him into the world, which human justice cannot do. If I, born a bloodthirsty and savage brute, inheriting these qualities from others, kill you, my fellow-men will very justly hang me, but I shall not be visited with the horrible remorse which would be my real punishment if, my nature being higher, I had done the same thing.

The absolute justice of the system of things is as clear to me as any scientific fact. The gravitation of sin to sorrow is as certain as that of the earth to the sun, and more so—for experimental proof of the fact is within reach of us all—nay, is before us all in our own lives, if we had but the eyes to see it.

Not only, then, do I disbelieve in the need for compensation, but I believe that the seeking for rewards and punishments out of this life leads men to a ruinous ignorance of the fact that their inevitable rewards and punishments are here.

If the expectation of hell hereafter can keep me from evil-doing, surely *a fortiori* the certainty of hell now will do so? If a man could be firmly impressed with the

belief that stealing damaged him as much as swallowing
arsenic would do (and it does), would not the dissuasive
force of that belief be greater than that of any based on
mere future expectations?

And this leads me to my other point.

As I stood behind the coffin of my little son the other
day, with my mind bent on anything but disputation,
the officiating minister read, as a part of his duty, the
words, "If the dead rise not again, let us eat and drink,
for to-morrow we die." I cannot tell you how inexpres-
sibly they shocked me.   Paul had neither wife nor child,
or he must have known that his alternative involved a
blasphemy against all that was best and noblest in
human nature.   I could have laughed with scorn.
What! because I am face to face with irreparable loss,
because I have given back to the source from whence it
came, the cause of a great happiness, still retaining through
all my life the blessings which have sprung and will
spring from that cause, I am to renounce my manhood,
and, howling, grovel in bestiality?   Why, the very apes
know better, and if you shoot their young, the poor
brutes grieve their grief out and do not immediately seek
distraction in a gorge.

Kicked into the world a boy without guide or training,
or with worse than none, I confess to my shame that few
men have drunk deeper of all kinds of sin than I.
Happily, my course was arrested in time—before I had
earned absolute destruction—and for long years I have
been slowly and painfully climbing, with many a fall,
towards better things.   And when I look back, what do
I find to have been the agents of my redemption?   The
hope of immortality or of future reward?   I can honestly
say that for these fourteen years such a consideration has
not entered my head.   No, I can tell you exactly what
has been at work.   *Sartor Resartus* led me to know that
a deep sense of religion was compatible with the entire
absence of theology.   Secondly, science and her methods

gave me a resting-place independent of authority and tradition. Thirdly, love opened up to me a view of the sanctity of human nature, and impressed me with a deep sense of responsibility.

If at this moment I am not a worn-out, debauched, useless carcass of a man, if it has been or will be my fate to advance the cause of science, if I feel that I have a shadow of a claim on the love of those about me, if in the supreme moment when I looked down into my boy's grave my sorrow was full of submission and without bitterness, it is because these agencies have worked upon me, and not because I have ever cared whether my poor personality shall remain distinct for ever from the All from whence it came and whither it goes.

And thus, my dear Kingsley, you will understand what my position is. I may be quite wrong, and in that case I know I shall have to pay the penalty for being wrong. But I can only say with Luther, "Gott helfe mir, Ich kann nichts anders."

I know right well that 99 out of 100 of my fellows would call me atheist, infidel, and all the other usual hard names. As our laws stand, if the lowest thief steals my coat, my evidence (my opinions being known) would not be received against him.[1]

But I cannot help it. One thing people shall not call me with justice, and that is—a liar. As you say of yourself, I too feel that I lack courage; but if ever the occasion arises when I am bound to speak, I will not shame my boy.

I have spoken more openly and distinctly to you than I ever have to any human being except my wife.

If you can show me that I err in premises or conclusion, I am ready to give up these as I would any other theories. But at any rate you will do me the justice to believe that I have not reached my conclusions without

---

[1] The law with respect to oaths was reformed in 1869.

the care befitting the momentous nature of the problems involved.

And I write this the more readily to you, because it is clear to me that if that great and powerful instrument for good or evil, the Church of England, is to be saved from being shivered into fragments by the advancing tide of science—an event I should be very sorry to witness, but which will infallibly occur if men like Samuel of Oxford are to have the guidance of her destinies—it must be by the efforts of men who, like yourself, see your way to the combination of the practice of the Church with the spirit of science. Understand that all the younger men of science whom I know intimately are *essentially* of my way of thinking. (I know not a scoffer or an irreligious or an immoral man among them, but they all regard orthodoxy as you do Brahmanism.) Understand that this new school of the prophets is the only one that can work miracles, the only one that can constantly appeal to nature for evidence that it is right, and you will comprehend that it is of no use to try to barricade us with shovel hats and aprons, or to talk about our doctrines being "shocking."

I don't profess to understand the logic of yourself, Maurice, and the rest of your school, but I have always said I would swear by your truthfulness and sincerity, and that good must come of your efforts. The more plain this was to me, however, the more obvious the necessity to let you see where the men of science are driving, and it has often been in my mind to write to you before.

If I have spoken too plainly anywhere, or too abruptly, pardon me, and do the like to me.

My wife thanks you very much for your volume of sermons.—Ever yours very faithfully,

T. H. HUXLEY.

A letter written in reply to the suggestion that he

should carry out Hooker's own good resolutions of
keeping out of the turmoil of life, and devoting himself to pure science, seems to indicate in its tone
something of the stress of the time when it was
written—

JERMYN STREET, *December* 19, 1860.

MY DEAR HOOKER—What with one thing and another, I have almost forgotten to answer your note—and
first, as to the business matter. . . . Next as to my
own private affairs, the youngster is "a-swelling wisibly,"
and my wife is getting on better than I hoped, though
not quite so well as I could have wished. The boy's
advent is a great blessing to her in all ways. For myself I hardly know yet whether it is pleasure or pain.
The ground has gone from under my feet once, and I
hardly know how to rest on anything again. Irrational,
you will say, but nevertheless natural. And finally as
to your resolutions, my holy pilgrim, they will be kept
about as long as the resolutions of other anchorites who
are thrown into the busy world, or I won't say that, for
assuredly you will take the world "as coolly as you can,"
and so shall I. But that coolness amounts to the red
heat of properly constructed mortals.

It is no use having any false modesty about the matter.
You and I, if we last ten years longer, and you by à
long while first, will be the representatives of our respective lines in this country. In that capacity we shall
have certain duties to perform to ourselves, to the outside world, and to science. We shall have to swallow praise
which is no great pleasure, and to stand multitudinous
bastings and irritations, which will involve a good deal of
unquestionable pain. Don't flatter yourself that there is
any moral chloroform by which either you or I can render
ourselves insensible or acquire the habit of doing things
coolly. It is assuredly of no great use to tear one's self

to pieces before one is fifty. But the alternative, for
men constructed on the high pressure tubular boiler
principle, like ourselves, is to lie still and let the devil
have his own way. And I will be torn to pieces before
I am forty sooner than see that.

I have been privately trading on my misfortunes in
order to get a little peace and quietness for a few months.
If I can help it, I don't mean to do any dining-out this
winter, and I have cut down Societies to the minimum of
the Geological, from which I cannot get away.

But it won't do to keep this up too long. By and by
one must drift into the stream again, and then there is
nothing for it but to pull like mad unless we want to be
run down by every collier.

I am going to do one sensible thing, however, viz. to
rush down to Llanberis with Busk between Christmas
Day and New Year's Day and get my lungs full of hill-
air for the coming session.

I was at Down on Saturday and saw Darwin. He
seems fairly well, and his daughter was up and looks
better than I expected to see her.—Ever yours faithfully,

T. H. HUXLEY.

Meanwhile, he took the opportunity to make the
child's birth a new link with his old friend. His
wife was desirous of having the boy christened; but
though, with a feeling which in part corresponded to
Descartes' *morale par provision*, he afterwards was not
unwilling to regard the ceremony as providing a link
with the official spiritual organisation of its country
which a child could either ignore or continue on
reaching intellectual maturity, still at the time he
was vexed and uneasy at having to assist at a rite
which to him was void of meaning. The only way
of turning it into a reality, he exclaims, is by

making it an extra bond with one's friends. And so, being himself ready to take up the relationship to a friend's child so long as he was not called upon to utter a declaration of belief in tenets he did not hold, he writes to Hooker :—

You know my opinions on these matters, and I would not ask you to do anything I would not do myself, so if you consent——you shall be asked to do nothing else than to help devour the christening feed, and be as good a friend to the boy as you have been to his father.

14 WAVERLEY PLACE, Jan. 6, 1861.

MY DEAR HOOKER—My wife and I were very pleased to get your hearty and kind acceptance of Godfathership. We shall not call upon you for some time, I fancy, as the mistress doesn't get strong very fast. However, I am only glad she is as well as she is. She came down yesterday for the first time.

It is very pleasant to get such expressions of opinion as I have had from you, Lyell, and Darwin about the *Review.* They make me quite hopeful about its prosperity, as I am sure we shall be able to do better than our first number.

I am glad you liked what I said in the opening of my article.[1] I wish not to be in any way confounded with

---

[1] In the *Natural History Review* (1861, p. 67).—"The proof of his claim to independent parentage will not change the brutishness of man's lower nature ; nor, except in those valet souls who cannot see greatness in their fellow because his father was a cobbler, will the demonstration of a pithecoid pedigree one whit diminish man's divine right of kingship over nature ; nor lower the great and princely dignity of perfect manhood, which is an order of nobility not inherited, but to be won by each of us, so far as he consciously seeks good and avoids evil, and puts the faculties with which he is endowed to their fittest use."

the cynics who delight in degrading man, or with the common run of materialists, who think mind is any the lower for being a function of matter. I dislike them even more than I do the pietists.

Some of these days I shall look up the ape question again, and go over the rest of the organisation in the same way. But in order to get a thorough grip of the question, I must examine into a good many points for myself. The results, when they do come out, will, I foresee, astonish the natives.

I am cold-proof, and all the better for the Welsh trip. To say truth, I was just on the edge of breaking down when I went. Did I ever send you a letter of mine on the teaching of Natural History? It was published while you were away, and I forget whether I sent it or not. However, a copy accompanies this note . . .

Of course there will be room for your review and welcome. I have put it down and reckon on it.—Ever yours faithfully,

T. H. HUXLEY.

Huxley returned from the trip to Wales in time to be with his wife for the New Year. The plot she had made with Dr. Tyndall had been entirely successful. The threatened breakdown was averted. Wales in winter was as good as Switzerland. Of the ascent of Snowdon he writes on December 28: "Both Tyndall and I voted it under present circumstances as good as most things Alpine."

His wife, however, continued in very weak health. She was prostrated by the loss of her little boy. So in the middle of March he gladly accepted Mr. Darwin's invitation for her and the three children to spend a fortnight in the quiet of his house at

Down, where he himself managed to run down for a week-end. "It appears to me," he writes to his wife, "that you are subjecting poor Darwin to a savage Tennysonian persecution. I shall see him looking like a martyr and have to talk double science next Sunday."

In April another good friend, Dr. Bence Jones, lent the invalid his house at Folkestone for three months. Unable even to walk when she went there, her recovery was a slow business. Huxley ran down every week; his brother George and his wife also were frequent visitors. Meanwhile he resolved to move into a new house, in order that she might not return to a place so full of sorrowful memories. On May 30 he effected the move to a larger house not half a mile away from Waverley Place —26 Abbey Place (now 23 Abercorn Place). Here also Mrs. Heathorn lived for the next year, my grandfather, over seventy as he was, being compelled to go out again to Australia to look after a business venture of his which had come to grief.

Meantime the old house was still on his hands for another year. Trying to find a tenant, he writes on May 21, 1861 :—

I met J. Tyndall at Ramsay's last night, and I think he is greatly inclined to have the house. I gave him your message and found that a sneaking kindness for the old house actuated him a good deal in wishing to take it. It is not a bad fellow, and we won't do him much on the fixtures.

Eventually Tyndall and his friend Hirst established themselves there.

This spring Professor Henslow, Mrs. Hooker's father, a botanist of the first rank, and a man extraordinarily beloved by all who came in contact with him, was seized with a mortal illness, and lingered on without hope of recovery through almost the whole of April. Huxley writes :—

JERMYN STREET, *April* 4, 1861.

MY DEAR HOOKER—I am very much grieved and shocked by your letter. The evening before last I heard from Busk that your father-in-law had been ill, and that you had been to see him, and I meant to have written to you yesterday to inquire, but it was driven out of my head by people coming here. And then I had a sort of unreasonable notion that I should see you at the Linnæan Council to-day and hear that all was right again. God knows, I feel for you and your poor wife. Knowing what a great rift the loss of a mere undeveloped child will leave in one's life, I can faintly picture to myself the great and irreparable vacuity in a family circle caused by the vanishing out of it of such a man as Henslow, with great acquirements, and that great calm catholic judgment and sense which always seemed to me more prominent in him than in any man I ever knew.

He had intellect to comprehend his highest duty distinctly, and force of character to do it ; which of us dare ask for a higher summary of his life than that ? For such a man there can be no fear in facing the great unknown, his life has been one long experience of the substantial justice of the laws by which this world is governed, and he will calmly trust to them still as he lays his head down for his long sleep.

You know all these things as well as I do, and I know as well as you do that such thoughts do not cure heartache or assuage grief. Such maladies, when men are as old as you and I are, are apt to hang about one a long time, but I find that if they are faced and accepted as part of our fair share of life, a great deal of good is to be got out of them. You will find that too, but in the meanwhile don't go and break yourself down with over wear and tear. The heaviest pull comes after the excitement of a catastrophe of this kind is over.

Believe in my affectionate sympathy with you, and that I am, my dear old fellow, yours ever,

T. H. Huxley.

And again on the 18th :—

Many thanks for your two letters. It would be sad to hear of life dragging itself out so painfully and slowly, if it were not for what you tell me of the calmness and wisdom with which the poor sufferer uses such strength as is left him.

One can express neither wish nor hope in such a case. With such a man what is will be well. All I have to repeat is, don't knock yourself up. I wish to God I could help you in some way or other beyond repeating the parrot cry. If I can, of course you will let me know.

In June 1861 a jotting in his notebook records that he is at work on the chick's skull, part of the embryological work which he took up vigorously at this time, and at once the continuation of his researches on the Vertebrate Skull, embodied in his Croonian lecture of 1858, and the beginning of a long series of investigations into the structure of birds. There is a reference to this in a very interesting letter dealing chiefly with what he conceived to be the cardinal point of the Darwinian theory :—

26 ABBEY PLACE, *Sept.* 4, 1861.

MY DEAR HOOKER—Yesterday being the first day I went to the Athenæum after reading your note, I had a look at, and a good laugh over, the *Quarterly* article. Who can be the writer?

I have been so busy studying chicken development, a difficult subject to which I had long ago made up my mind to devote my first spare time, that I have written you no word about your article in the *Gardener's Chronicle.* I quite agree with the general tendency of your argument, though it seems to me that you put your view rather too strongly when you seem to question the position "that, as a rule, resemblances prevail over differences" between parent and offspring.   Surely, as a rule, resemblances *do* prevail over differences, though I quite agree with you that the latter have been far too much overlooked.   The great desideratum for the species question at present seems to me to be the determination of the law of variation.   Because no law has yet been made out, Darwin is obliged to speak of variation as if it were spontaneous or a matter of chance, so that the bishops and superior clergy generally (the only real atheists and believers in chance left in the world) gird at him as if he were another Lucretius.

It is [in] the recognition of a tendency to variation apart from the variation of what are ordinarily understood as external conditions that Darwin's view is such an advance on Lamarck.   Why does not somebody go to work experimentally, and get at the law of variation for some one species of plant?

What a capital article that was in the *Athenæum* the

other day *apud* the Schlagintweits.[1]   Don Roderigo [2] is very wroth at being made responsible with Sabine, and indeed I think he had little enough to do with it.

You will see a letter from him in this week's *Athenæum.*—Ever yours faithfully,   T. H. HUXLEY.

---

[1] The brothers Schlagintweit (four of whom were ultimately employed), who had gained some reputation for their work on the Physical Geography of the Alps, were, on Humboldt's recommendation, despatched by the East India Company in 1854-55-56, to the Deccan, and especially to the Himalayan region (where they were the first Europeans to cross the Kuenlun Mountains), in order to correlate the instruments and observations of the several magnetic surveys of India.   But they enlarged the scope of their mission by professing to correct the great trigonometrical survey, while the contract with them was so loosely drawn up that they had practically a roving commission in science, to make researches and publish the results—up to nine volumes—in all manner of subjects, which in fact ranged from the surveying work to ethnology, and were crowned by an additional volume on Buddhism !   The original cost to the Indian Government was estimated at £15,000 ; the allowances from the English Government during the inordinately prolonged period of arranging and publishing materials, including payment for sixty copies of each volume, atlas, and so forth, as well as personal payments, came to as much more.

Unfortunately the results were of less value than was expected. The attempt to correct the work done with the large instruments of the trigonometrical survey by means of far smaller instruments was absurd ; away from the ground covered by the great survey the figures proved to be very inaccurate.   The most annoying part of the affair was that it absorbed the State aid which might have been given to more valuable researches.

The Council of the Royal Society had been consulted as to the advisability of despatching this expedition and opposed it, for there were in the service of the Company not a few men admirably qualified for the duty, whose scientific services had received scant appreciation.   Nevertheless, the expedition started after all, with the approval of Colonel Sabine, the president.   In the last months of 1866, Huxley drew up for the Royal Society a report upon the scientific value of the results of the expedition.

[2] Sir Roderick Murchison.

CHAPTER XVII

1861-1863

It has been seen that the addition of journalistic work in science to the mass of original research and teaching work upon which Huxley was engaged, called forth a remonstrance from both Lyell and Darwin. To Hooker it seemed still more serious that he was dividing his allegiance, and going far afield in philosophy, instead of concentrating himself upon natural science. He writes :—

I am sorry to hear that you are so poorly, and wish I could help you to sit down and work quietly at pure science. You have got into a whirlpool, and should strike out vigorously at the proper angle, not attempt to breast the whole force of the current, nor yet give in to it. Do take the counsel of a quiet looker-on and withdraw to your books and studies in pure Natural History ; let modes of thought alone. You may make a very good naturalist, or a very good metaphysician (of that I know nothing, don't despise me), but you have neither time nor place for both.

However, it must be remarked that this love of philosophy, not recently acquired either, was only

part of the passion for general principles underlying
the facts of science which had always possessed him.
And the time expended upon it was not directly
taken from the hours of scientific work; he would
read in bed through the small hours of the night,
when sleep was slow in coming to him.   In this way
he got through an immense amount of philosophy in
the course of several years.   Not that he could "state
the views of so and so " upon any given question, or
desired such kind of knowledge; he wished to find
out and compare with his own the answers which
other thinkers gave to the problems which interested
himself.

A gentler reproof of this time touches his hand-
writing, which was never of the most legible, so that
his foreign correspondents in particular sometimes
complained.   Haeckel used to get his difficulties
deciphered by his colleague Gegenbaur.   I cannot
forbear quoting the delicate remonstrance of Professor
Lacaze Duthiers, and the flattering remedy he proposed :—

March 14.—Je lis l'Anglais imprimé, mais vos écritures
anglaises sont si rapides, qu'il m'est quelquefois difficile de
m'en sortir.   On me dit que vous écrivez si bien le
français que je crois que je vous lirais bien mieux dans ma
langue !

On his return from examining at Dublin, he again
looked over proofs on biological subjects for Mr.
Spencer.

JERMYN STREET, *Aug.* 3, 1861.

MY DEAR SPENCER—I have been absent on a journey to Dublin and elsewhere [1] nearly all this week, and hence your note and proof did not reach me till yesterday. I have but just had time to glance through the latter, and I need hardly say how heartily I concur in its general tenor. I have, however, marked one or two passages which I think require some qualification. Then, at p. 272, the fact that the vital manifestations of plants depend as entirely as those of animals upon the fall towards stable equilibrium of the elements of a complex protein compound is not sufficiently prominent. It is not so much that plants are deoxidisers and animals oxidisers, as that plants are manufacturers and animals consumers. It is true that plants manufacture a good deal of non-nitrogenous produce in proportion to the nitrogenous, but it is the latter which is chiefly useful to the animal consumer and not the former. This point is a very important one, which I have never seen clearly and distinctly put—the prettiness of Dumas' circulation of the elements having seduced everybody.

Of course this in no way affects the principle of what you say. The statements which I have marked at p. 276 and 278 should have their authorities given, I think. I should hardly like to commit myself to them absolutely.

You will, if my memory does not mislead me, find authority for my note at p. 283 in Stephenson's life. I think old Geo. Stephenson brought out his views at breakfast at Sir R. Peel's when Buckland was there.

These are all the points that strike me, and I do not keep your proof longer (I send it by the same post as this note), because I fear you may be inconvenienced by the delay.

Tyndall is unfortunately gone to Switzerland, so that

---

[1] Visiting Sir Philip Egerton at Oulton Park.

I cannot get you his comments.   Whether he might have
picked holes in any detail or not I do not know, but I
know his opinions sufficiently well to make sure in his
agreement with the general argument.   In fact a favourite
problem of his is—Given the molecular forces in a mutton
chop, deduce Hamlet or Faust therefrom.   He is confident
that the Physics of the Future will solve this easily.

I am grieved to hear such a poor account of your
health ; I believe you will have to come at last to the
heroic remedy of matrimony, and if "gynopathy" were a
mode of treatment that could be left off if it did not suit
the constitution, I should decidedly recommend it.

But it's worse than opium-eating—once begin and you
must go on, and so, though I ascribe my own good
condition mainly to the care my wife takes of me, I dare
not recommend it to you, lest perchance you should get
hold of the wrong medicine.

Beyond spending a night awake now and then I am
in very good order, and I am going to spend my vacation
in a spasmodic effort to lick the *Manual* into shape and
work off some other arrears.

My wife is very fairly well, and, I trust, finally freed
from all the symptoms which alarmed me so much.   I
dread the coming round of September for her again, but it
must be faced.

The babbies are flourishing ; and beyond the facts
that we have a lunatic neighbour on one side and an
empty house on the other, that it has cost me about twice
as much to get into my house as I expected, that the
cistern began to leak and spoil a ceiling, and such other
small drawbacks, the new house is a decided success.

I forget whether I gave you the address, which is—

26 ABBEY PLACE,

ST. JOHN'S WOOD.

You had better direct to me there, as after the 10th

of this month I shall not be here for six weeks.—Ever
yours faithfully,   T. H. HUXLEY.

October shows an unusual entry in his diary; the
sacrifice of a working evening to hear Jenny Lind
sing. Fond though he was of music, as those may
remember who ever watched his face at the Sunday
evening gatherings in Marlborough Place in the later
seventies, when there was sure to be at least a little
good music or singing either from his daughters or
some of the guests, he seldom could spare the time
for concert-going or theatre-going, and the occasional
notes of his bachelor days, "to the opera with
Spencer," had ceased as his necessary occupations
grew more engrossing.

This year his friend Hooker moved to Kew to act
as second in command to his father, Sir William
Hooker, the director of the Botanical Gardens. This
move made meetings between the two friends, except
at clubs and societies, more difficult, and was one of
the immediate causes of the foundation of the *x* Club.
It is this move which is referred to in the following
letters; the "poor client" being the wife of an old
messmate of his on the *Rattlesnake :*—

JERMYN STREET, *Nov.* 17.

MY DEAR HOOKER—My wife wrote to yours yesterday,
the enclosed note explaining the kitchen-revolution which,
it seems, must delay our meeting. When she had done,
however, she did not know where to direct it, and I am
no wiser, so I send it to you.

It's a horrid nuisance and I have sworn a few, but

that will not cook the dinner, however much it may prepare me for being cooked elsewhere. To complete my disgust at things in general, my wife is regularly knocked up with dining out twice this week, though it was only in the quietest way. I shall have to lock her up altogether.

X—— has made a horrid mess of it, and I am sorry to say, from what I know of him, that I cannot doubt where the fault lies. The worst of it is that he has a wife and three children over here, left without a penny or any means of support. The poor woman wrote to me the other day, and when I went to see her I found her at the last shilling and contemplating the workhouse as her next step. She has brothers in Australia, and it appeared to me that the only way to do her any good was to get her out. She cannot starve there, and there will be more hope for her children than an English poor-house. I am going to see if the Emigration Commissioners will do anything for her, as of course it is desirable to cut down the cost of exportation to the smallest amount.

It is most lamentable that a man of so much ability should have so utterly damned himself as X—— has, but he is hopelessly Celtic.

I shall be at the Phil. Club next Thursday.—Ever yours faithfully,   T. H. HUXLEY.

14 WAVERLEY PLACE,
Monday morning [*Nov.* 1861].

MY DEAR HOOKER—The obstinate manner in which Mrs. Hooker and you go on refusing to give any address leads us to believe that you are dwelling peripatetically in a "Wan" with green door and brass knocker somewhere on Wormwood Scrubbs, and that "Kew" is only a blind. So you see I am obliged to inclose Mrs. Hooker's epistle to you.

You shall have your own way about the dinner, though we shall have triumphed over all domestic difficulties by that time, and the first lieutenant scorns the idea of being "worrited" about anything. I only grieve it is such a mortal long way for you to come.

I could find it in my heart to scold you well for your generous aid to my poor client. I assure you I told you all about the case because it was fresh in my mind, and without the least notion of going to you for that kind of aid. May it come back to you in some good shape or other.

I find it is no use to look for help from the emigration people, but I have no fear of being able to get the £50 which will send them out by the *Walter Hood*.

Would it be fair to apply to Bell in such a case? I will have a talk to you about it at the Phil. Club.—Ever, my dear Hooker, yours faithfully,   T. H. HUXLEY.

In 1862, in addition to all the work connected with the species question already detailed, Huxley published three paleontological papers,[1] while the paper on the "Anatomy and Development of Pyrosoma," first read on December 1, 1859, was now published in the *Proceedings of the Linnean Society*.

In the list of work in hand are four paleontological papers,[2] besides the slowly progressing *Manual of Comparative Anatomy*.

When he went north to deliver his lectures at Edinburgh "On the relation of Man to the Lower

---

[1] "On the new Labyrinthodonts from the coal-field of Edinburgh"; "On a Stalk-eyed Crustacean from the coal-fields of Paisley"; and "On the Teeth of Diprotodon."

[2] "On Indian Fossils," on "Cephalaspis and Pteraspis," on "Stagonolepis," and a "Memoir descriptive of Labyrinthodont remains from the Trias and Coal of Britain," which he first treated of in 1858, "clearly establishing for the first time the vertebrate nature of these remains."—Sir M. Foster, Obit. Notice, *Proc. R. S.* lix. 55.

Animals," he took the opportunity of examining fossils at Forfar, and lectured also at Glasgow; while at Easter he went to Ireland; on March 15 he was at Dublin, lecturing there on the 25th.

Reference has already been made (in the letter to C. Darwin of May 6, 1862) to the unsatisfactory state of Huxley's health. He was further crippled by neuralgic rheumatism in his arm and shoulder, and to get rid of this, went on July 1 to Switzerland for a month's holiday. Reaching Grindelwald on the 4th, he was joined on the 6th by Dr. Tyndall, and with him rambled on the glacier and made an expedition to the Faulhorn. On the 13th they went to the Rhone glacier, meeting Sir J. Lubbock on their way, at the other side of the Grimsel. Both here and at the Eggischhorn, where they went a few days later, Huxley confined himself to easy expeditions, or, as his notebook has it, stayed "quiet" or "idle," while the hale pair ascended the Galenstock and the Jungfrau.

By July 28 he was home again in time for an examiners' meeting at the London University the next day, and a *viva voce* in physiology on the 4th August, before going to Scotland to serve on the Fishery Commission.

This was the first of the numerous commissions on which he served. With his colleagues, Dr. Lyon Playfair (afterwards Lord Playfair) and Colonel Maxwell, he was busy from August 8 to September 16, chiefly on the west coast, taking evidence from

the trawlers and their opponents, and making direct
investigations into the habits of the herring.

The following letter to Mr. (afterwards Sir W. H.)
Flower, then Curator of the Royal College of Surgeons
Museum, refers to this trip and to his appointment
to the examinership in physiology at the College of
Surgeons, for which he had applied in May and which
he held until 1870.   Mr. Flower, indeed, was deeply
interested at this time in the same problems as
Huxley, and helped his investigations for *Man's Place*
by making a number of dissections to test the disputed
relations between the brain of man and of the apes.

HOTEL DE LA JUNGFRAU,
AEGGISCHHORN, *July* 18, 1862.

MY DEAR FLOWER—Many thanks for your letter.   I
shall make my acknowledgments to the council in due
form when I have read the official announcement on my
return to England.   I trust they will not have occasion
to repent declining Dr. ——'s offer.   At any rate I shall
do my best.

I am particularly obliged to you for telling me about
the Dijon bones.   Dijon lies quite in my way in returning to England, and I shall stop a day there for the
purpose of making the acquaintance of M. Nodet and his
Schizopleuron.   I have a sort of dim recollection that
there are some other remains of extinct South American
mammals in the Dijon Museum which I ought to see.

Your news about the lower jaw made me burst out
into such an exclamation that all the salle-à-manger
heard me !   I saw the fitness of the thing at once.   The
foramen and the shape of the condyle ought to have
suggested it at once.

I have had a very pleasant trip, passing through

Grindelwald, the Aar valley, and the Rhone valley, as far as here ; but, up to the day before yesterday, my health remained very unsatisfactory, and I was terribly teased by the neuralgia or rheumatism or whatever it is.

On that day, however, I had a very sharp climb involving a great deal of exertion and a most prodigious sweating, and on the next morning I really woke up a new man. Yesterday I repeated the dose, and I am in hopes now that I shall come back fit to grapple with all the work that lies before me.—Ever, my dear Flower, yours very faithfully,   T. H. HUXLEY.

This autumn he gladly took on what appeared to be an additional piece of work. On October 12 he writes from 26 Abbey Place :—

I saw Flower yesterday, and I find that my present colleague in the Hunterian Professorship wishes to get rid of his share in the lectures, having, I suppose, at the eleventh hour discovered his incompetency. It looks paradoxical to say so, but it will really be easier for me to give eighteen or twenty-four lectures than twelve, so that I have professed my readiness to take as much as he likes off his hands.

This professorship had been in existence for more than sixty years, for when the Museum of the famous anatomist John Hunter was entrusted to the College of Surgeons by the Government, the condition was made "that one course of lectures, not less than twenty-four in number, on comparative anatomy and other subjects, illustrated by the preparations, shall be given every year by some member of the company." Huxley arranged to publish from year to year the substance of his lectures on the vertebrates, " and by that process to bring out eventually a comprehensive, though condensed, systematic work on *Comparative Anatomy.*" [1]

Of the labour entailed in this course, the late Sir W. H. Flower wrote :—

When, in 1862, he was appointed to the Hunterian Professorship at the College of Surgeons, he took for the subject of several yearly courses of lectures the anatomy of the vertebrata, beginning with the primates, and as the subject was then rather new to him, and as it was a rule with him never to make a statement in a lecture which was not founded upon his own actual observation, he set to work to make a series of original dissections of all the forms he treated of. These were carried on in the workroom at the top of the college, and mostly in the evenings, after his daily occupation at Jermyn Street (the School of Mines, as it was then called) was over, an arrangement which my residence in the college buildings enabled me to make for him. These rooms contained a large store of material, entire or partially dissected animals preserved in spirit, which, unlike those mounted in the museum, were available for further investigation in any direction, and these, supplemented occasionally by fresh subjects from the Zoological Gardens, formed the foundation of the lectures. . . . On these evenings it was always my privilege to be with him, and to assist in the work in which he was engaged. In dissecting, as in everything else, he was a very rapid worker, going straight to the point he wished to ascertain with a firm and steady hand, never diverted into side issues, nor wasting any time in unnecessary polishing up for the sake of appearances ; the very opposite, in fact, to what is commonly known as " finikin." His great facility for

---

[1] *Comparative Anatomy,* vol. i. Preface.

bold and dashing sketching came in most usefully in this work, the notes he made being largely helped out with illustrations.

The following is the letter in which he makes himself known to Professor Haeckel of Jena, who, in his thanks for the specimens, bewails the lot of "us poor inland Germans, who have to get help from England."

THE ROYAL SCHOOL OF MINES,
JERMYN STREET, LONDON, *October*, 28, 1862.

Sir—A copy of your exceedingly valuable and beautiful monograph, "Die Radiolarien," came into my hands two or three days ago, and I have been devoting the little leisure I possess just at present to a careful study of its contents, which are to me profoundly interesting and instructive.

Permit me to say this much by way of introduction to a request which I have to prefer, which is, that you will be good enough to let me have a copy of your Habilitationsschrift, *De Rhizopodum Finibus*, if you have one to spare. If it is sent through Frommans of Jena to the care of Messrs. Williams and Norgate, London, it will reach me safely.

I observe that in your preface you state that you have no specimen of the famous Barbadoes deposit. As I happen to possess some from Schomburgk's own collection, I should be ashamed to allow you any longer to suffer from that want, and I beg your acceptance of the inclosed little packet. If this is not sufficient, pray let me know and I will send you as much more.

If you desire it, I can also send you some of the Oran earth, and as much as you like of the Atlantic deep-sea soundings, which are almost entirely made up of *Globigerina* and *Polycistina*.—I am, Sir, yours very faithfully,　　THOMAS H. HUXLEY.

The next letter refers to the scientific examinations at the University of London.

Dec. 4, 1862.

My DEAR HOOKER—I look upon you as art and part of the *Natural History Review*, though not ostensibly one of the gang, so I bid you to a feast, partly of reason and partly of mutton, at my house on December 11 (being this day week) at half-past six. Do come if you can, for we have not seen your ugly old phiz for ages, and should be comforted by an inspection thereof, however brief.

I did my best to get separate exhibitions for Chemistry, Botany, and Zoological Biology, at the committee yesterday,[1] and I suspect from your letter that if you had been there you would have backed me. However, it is clear they only mean to give separate exhibits for Chemistry and Biology as a whole.

Because Botany and Zoology are, philosophically speaking, cognate subjects, people are under the delusion that it is easier to work both up at the same time, than it would be to work up, say, Chemistry and Botany. Just fancy asking a young man who has heaps of other things to work up for the B.Sc., to qualify himself for honours both in [zoology and] botany, histological, systematic, and physiological. That is to say, to get a *practical knowledge* of both these groups of subjects.

I really think the botanical and zoological examiners ought to memorialise the senate jointly on the subject. The present system leads to mere sham and cram.—
Ever yours,   T. H. HUXLEY.

The year 1863, notable for the publication of Huxley's first book, found him plunged deep in an immense quantity of work of all sorts. He

[1] At the London University.

was still examiner in Physiology and Comparative Anatomy at the London University, a post he held from 1855 to 1863, and again from 1865 to 1870, "making," as Sir Michael Foster says, "even an examination feel the influence of the new spirit in biology; and among his examinees at that time there was one at least who, knowing Huxley by his writings, but by his writings only, looked forward to the *viva voce* test, not as a trial, but as an occasion of delight."

In addition to the work mentioned in the following letters, I note three lectures at Hull on April 6, 8, and 10; a paper on "Craniology" (January 17), and his "Letter on the Human Remains in the Shell Mounds," in the *Ethnological Society's Transactions*, while the Fishery Commission claimed much of his time, either at the Board of Trade, or travelling over the north, east, and south coasts from the end of July to the beginning of October, and again in November and December.

JERMYN STREET, *April* 30, 1863.

MY DEAR KINGSLEY—I am exceedingly pleased to have your good word about the lectures,[1]—and I think I shall thereby be encouraged to do what a great many people have wished—that is, to bring out an enlarged and revised edition of them.

The only difficulty is time—if one could but work five-and-twenty hours a day !

With respect to the sterility question, I do not think there is much doubt as to the effect of breeding in and

---

[1] See p. 300.

in in destroying fertility. But the sterility which must
be obtained by the selective breeder in order to convert
his morphological species into physiological species—
such as we have in nature—must be quite irrespective
of breeding in and in.

There is no question of breeding in and in between a
horse and an ass, and yet their produce is usually a sterile
hybrid.

So if Carrier and Tumbler, *e.g.*, were physiological
species equivalent to Horse and Ass, their progeny ought
to be sterile or semi-sterile. So far as experience has
gone, on the contrary, it is perfectly fertile—as fertile
as the progeny of Carrier and Carrier or Tumbler and
Tumbler.

From the first time that I wrote about Darwin's book
in the *Times* and in the *Westminster* until now, it has
been obvious to me that this is the weak point of Darwin's
doctrine. He *has* shown that selective breeding is a *vera
causa* for morphological species ; he has not yet shown it
a *vera causa* for physiological species.

But I entertain little doubt that a carefully devised
system of experimentation would produce physiological
species by selection—only the feat has not been performed
yet.

I hope you received a copy of *Man's Place in Nature*,
which I desired should be sent to you long ago. Don't
suppose I ever expect an acknowledgment of a book—it
is one of the greatest nuisances in the world to have that
to do, and I never do it—but as you mentioned the
Lectures and not the other, I thought it might not have
reached you. If it has not, pray let me know and a copy
shall be forwarded, as I want you very much to read
Essay No. 2.

I have a great respect for all the old bottles, and if
the new wine can be got to go into them and not burst
them I shall be very glad—I confess I do not see my
way to it ; on the contrary, the longer I live and the

more I learn the more hopeless to my mind becomes the contradiction between the theory of the universe as understood and expounded by Jewish and Christian theologians, and the theory of the universe which is every day and every year growing out of the application of scientific methods to its phenomena.

Whether astronomy and geology can or cannot be made to agree with the statements as to the matters of fact laid down in Genesis—whether the Gospels are historically true or not—are matters of comparatively small moment in the face of the impassable gulf between the anthropomorphism (however refined) of theology and the passionless impersonality of the unknown and unknowable which science shows everywhere underlying the thin veil of phenomena.

Here seems to me to be the great gulf fixed between science and theology—beside which all Colenso controversies, reconcilements of Scripture à la Pye Smith, etc., cut a very small figure.

You must have thought over all this long ago ; but steeped as I am in scientific thought from morning till night, the contrast has perhaps a greater vividness to me. I go into society, and except among two or three of my scientific colleagues I find myself alone on these subjects, and as hopelessly at variance with the majority of my fellow-men as they would be with their neighbours if they were set down among the Ashantees. I don't like this state of things for myself—least of all do I see how it will work out for my children. But as my mind is constituted, there is no way out of it, and I can only envy you if you can see things differently.—Ever yours very faithfully,   T. H. HUXLEY.

JERMYN STREET, *May* 5, 1863.

MY DEAR KINGSLEY—My wife and children are away at Felixstow on the Suffolk coast, and as I run down on Saturday and come back on Monday your MS. has been

kept longer than it should have been.   I am quite agreed
with the general tenor of your argument; and indeed I
have often argued against those who maintain the intellectual gulf between man and the lower animals to be an
impassable one, by pointing to the immense intellectual
chasm as compared to the structural differences between
two species of bees or between sheep and goat or dog and
wolf. So again your remarks upon the argument drawn
from the apparent absence of progression in animals seem
to me to be quite just. You might strengthen them
much by reference to the absence of progression in many
races of men. The West African savage, as the old
voyagers show, was in just the same condition two
hundred years ago as now—and I suspect that the
modern Patagonian is as nearly as possible the un-
improved representative of the makers of the flint
implements of Abbeville.

Lyell's phrase is very good, but it is a simple application of Darwin's views to human history. The advance
of mankind has everywhere depended on the production
of men of genius; and that production is a case of
"spontaneous variation" becoming hereditary, not by
physical propagation, but by the help of language, letters
and the printing press. Newton was to all intents and
purposes a "sport" of a dull agricultural stock, and his
intellectual powers are to a certain extent propagated by
the grafting of the "Principia," his brain-shoot, on us.

Many thanks for your letter. It is a great pleasure
to me to be able to speak out to any one who, like yourself, is striving to get at truth through a region of intellectual and moral influences so entirely distinct from
those to which I am exposed.

I am not much given to open my heart to anybody,
and on looking back I am often astonished at the way in
which I threw myself and my troubles at your head, in
those bitter days when my poor boy died. But the way
in which you received my heathen letters set up a free-

masonry between us, at any rate on my side ; and if they make you a bishop I advise you not to let your private secretary open any letters with my name in the corner, for they are as likely as not to contain matters which will make the clerical hair stand on end.

I am too much a believer in Butler and in the great principle of the "Analogy" that "there is no absurdity in theology so great that you cannot parallel it by a greater absurdity of Nature" (it is not commonly stated in this way), to have any difficulties about miracles.   I have never had the least sympathy with the *a priori* reasons against orthodoxy, and I have by nature and disposition the greatest possible antipathy to all the atheistic and infidel school.

Nevertheless, I know that I am, in spite of myself, exactly what the Christian world call, and, so far as I can see, are justified in calling, atheist and infidel.[1]   I cannot see one shadow or tittle of evidence that the great unknown underlying the phenomena of the universe stands to us in the relation of a Father—loves us and cares for us as Christianity asserts.   On the contrary, the whole teaching of experience seems to me to show that while the governance (if I may use the term) of the universe is rigorously just and substantially kind and beneficent, there is no more relation of affection between governor and governed than between me and the twelve judges.   I know the administrators of the law desire to do their best for everybody, and that they would rather not hurt me than otherwise, but I also know that under certain circumstances they will most assuredly hang me ; and that in any case it would be absurd to suppose them guided by any particular affection for me.

This seems to me to be the relation which exists between the cause of the phenomena of this universe and myself.   I submit to it with implicit obedience and

---

[1] But see vol. iii. chap. v.

perfect cheerfulness, and the more because my small intelligence does not see how any other arrangement could possibly be got to work as the world is constituted.

But this is what the Christian world calls atheism, and because all my toil and pains does not enable me to see my way to any other conclusion than this, a Christian judge would (if he knew it) refuse to take my evidence in a court of justice against that of a Christian ticket-of-leave man.

So with regard to the other great Christian dogmas, the immortality of the soul, and the future state of rewards and punishments, what possible objection *a priori* can I—who am compelled perforce to believe in the immortality of what we call Matter and Force and in a very unmistakable *present* state of rewards and punishments for all our deeds—have to these doctrines ? Give me a scintilla of evidence, and I am ready to jump at them.

But read Butler, and see to what drivel even his great mind descends when he has to talk about the immortality of the soul ! I have never seen an argument on that subject which from a scientific point of view is worth the paper it is written upon. All resolve themselves into this formula :—The doctrine of the immortality of the soul is very pleasant and very useful, therefore it is true.

All the grand language about "human aspiration," "consistency with the divine justice," etc. etc., collapses into this at last—Better the misery of the "Vale! in æternum vale !" ten times over than the opium of such empty sophisms—I have drunk of that cup to the bottom.

I am called away and must close my letter. Don't trouble to answer it unless you are so minded.—Ever yours faithfully,   T. H. HUXLEY.

JERMYN STREET, *May* 22, 1863.

MY DEAR KINGSLEY—Pray excuse my delay in replying to your letter. I have been very much pressed for time for these two or three days.

First touching the action of the spermatozoon. The best information you can find on the subject is, I think, in Newport's papers in the *Philosophical Transactions* for 1851, 1853, and 1854, especially the 1853 paper. Newport treats only of the Frog, but the information he gives is very full and definite. Allen Thomson's very accurate and learned article "Ovum" in Todd's *Cyclopædia* is also well worth looking through, though unfortunately it is least full just where you want most information. In French there is Coste's *Développement des Corps organisés* and the volume on "Development" by Bischoff in the French translation of the last edition of Soemmering's *Anatomy*.

So much for your inquiries as to the matters of fact. Next, as to questions of speculation. If any expression of ignorance on my part will bring us nearer we are likely to come into absolute contact, for the possibilities of "*may be*" are, to me, infinite.

I know nothing of Necessity, abominate the word Law (except as meaning that we know nothing to the contrary), and am quite ready to admit that there may be some place, "other side of nowhere," *par exemple*, where $2 + 2 = 5$, and all bodies naturally repel one another instead of gravitating together.

I don't know whether Matter is anything distinct from Force. I don't know that atoms are anything but pure myths. *Cogito, ergo sum* is to my mind a ridiculous piece of bad logic, all I can say at any time being "Cogito." The Latin form I hold to be preferable to the English "I think," because the latter asserts the existence of an Ego—about which the bundle of phenomena at present addressing you knows nothing. In fact, if I am pushed, metaphysical speculation lands me exactly where your friend Raphael was when his bitch pupped. In other words, I believe in Hamilton, Mansell, and Herbert Spencer so long as they are destructive, and I laugh at their beards as soon as they try to spin their own cobwebs.

Is this basis of ignorance broad enough for you ? If you, theologian, can find as firm footing as I, man of science, do on this foundation of minus nought—there will be nought to fear for our ever diverging.

For you see I am quite as ready to admit your doctrine that souls secrete bodies as I am the opposite one that bodies secrete souls—simply because I deny the possibility of obtaining any evidence as to the truth and falsehood of either hypothesis. My fundamental axiom of speculative philosophy is that *materialism and spiritualism are opposite poles of the same absurdity*—the absurdity of imagining that we know anything about either spirit or matter.

Cabanis and Berkeley (I speak of them simply as types of schools) are both asses, the only difference being that one is a black donkey and the other a white one.

This universe is, I conceive, like to a great game being played out, and we poor mortals are allowed to take a hand. By great good fortune the wiser among us have made out some few of the rules of the game, as at present played. We call them "Laws of Nature," and honour them because we find that if we obey them we win something for our pains. The cards are our theories and hypotheses, the tricks our experimental verifications. But what sane man would endeavour to solve this problem : given the rules of a game and the winnings, to find whether the cards are made of pasteboard or gold-leaf ? Yet the problem of the metaphysicians is to my mind no saner.

If you tell me that an Ape differs from a Man because the latter has a soul and the ape has not, I can only say it may be so ; but I should uncommonly like to know how [you know] either that the ape has not one or that the man has.

And until you satisfy me as to the soundness of your method of investigation, I must adhere to what seems to my mind a simpler form of notation—*i.e.* to suppose that all phenomena have the same substratum (if they have any), and that soul and body, or mental and physical phenomena, are merely diverse manifestations of that hypothetical substratum.   In this way, it seems to me, I obey the rule which works so well in practice, of always making the simplest possible suppositions.

On the other hand, if you are of a different opinion, and find it more convenient to call the x which underlies (hypothetically) mental phenomena, Soul, and the x which underlies (hypothetically) physical phenomena, Body, well and good.   The two-fluid theory and the one-fluid theory of electricity both accounted for the phenomena up to a certain extent, and both were probably wrong.   So it may be with the theories that there is only one x in nature or two x's or three x's.

For, if you will think upon it, there are only four possible ontological hypotheses now that Polytheism is dead.

I. There is no x = Atheism on Berkeleyan principles.

II. There is only one x = Materialism or Pantheism, according as you turn it heads or tails.

III. There are two x's } = Speculators *incertæ sedis.*
     Spirit and Matter }

IV. There are three x's } = Orthodox Theologians.
    God, Souls, Matter }

To say that I adopt any one of those hypotheses, as a representation of fact, would to my mind be absurd ; but No. 2 is the one I can work with best.   To return to my metaphor, it chimes in better with the rules of the game of nature than any other of the four possibilities, to my mind.

But who knows when the great Banker may sweep away table and cards and all, and set us learning a new game ?   What will become of all my poor counters then ?

It may turn out that I am quite wrong, and that there are no x's or 20 x's.

I am glad you appreciate the rich absurdities of the new doctrine of spontogenesis [?]. Against the doctrine of spontaneous generation in the abstract I have nothing to say. Indeed it is a necessary corollary from Darwin's views if legitimately carried out, and I think Owen smites him (Darwin) fairly for taking refuge in "Pentateuchal" phraseology when he ought to have done one of two things—(a) give up the problem, (b) admit the necessity of spontaneous generation. It is the very passage in Darwin's book to which, as he knows right well, I have always strongly objected. The x of science and the x of Genesis are two different x's, and for any sake don't let us confuse them together. Maurice has sent me his book. I have read it, but I find myself utterly at a loss to comprehend his point of view.—Ever yours faithfully,

T. H. HUXLEY.

The following letter is interesting, as showing his continued interest in the question of skull structure, as well as his relation to his friend and fellow-worker, Dr. W. K. Parker.[1]

JERMYN STREET, *March* 18, 1863.

MY DEAR PARKER·— Any conclusion that I have reached will seem to me all the better based for knowing that you have been near or at it, and I am therefore right glad to have your letter. If I had only time, nothing would delight me more than to go over your preparations, but these Hunterian Lectures are about the hardest bit of work I ever took in hand, and I am obliged to give every minute to them.

By and by I will gladly go with you over your vast material.

---

[1] William Kitchin Parker, 1823-1890, a medical practitioner and anatomist, Hunterian Professor of Comparative Anatomy at the Royal College of Surgeons, 1873.

Did you not some time ago tell me that you considered the Y-shaped bone (so-called presphenoid) in the Pike to be the true basisphenoid? If so, let me know before lecture to-morrow, that I may not commit theft unawares.

I have arrived at that conclusion myself from the anatomical relations of the bone in question to the brain and nerves.

I look upon the proposition opisthotis = turtle's "occipital externe" = Perch's Rocher (Cuvier) as the one thing needful to clear up the unity of structure of the bony cranium; and it shall be counted unto me as a great sin if I have helped to keep you back from it. The thing has been dawning upon me ever since I read Kölliker's book two summers ago, but I have never had time to work it out.—Ever yours faithfully,

T. H. HUXLEY.

The following extracts from a letter to Hooker and a letter to Darwin describe the pressure of his work at this time.

1863.

MY DEAR HOOKER— . . . I would willingly send a paper to the Linnean this year if I could, but I do not see how it is practicable. I lecture five times a week from now till the middle of February. I then have to give eighteen lectures at the Coll. Surgeons—six on classification, and twelve on the vertebrate skeleton. I must write a paper on this great new Glyptodon, with some eighteen to twenty plates. A preliminary notice has already gone to the Royal Society. I have a decade of fossil fish in progress; a fellow in the country *will* keep on sending me splendid new Labyrinthodons from the coal, and that d—d manual must come out.—*Ayez pitié de moi.* T. H. H.

JERMYN STREET, *July* 2, 1863.

MY DEAR DARWIN—I am horribly loth to say that I cannot do anything you want done; and partly for that reason and partly because we have been very busy here with some new arrangements during the last day or two, I did not at once reply to your note.

I am afraid, however, I cannot undertake any sort of new work. In spite of working like a horse (or if you prefer it, like an ass), I find myself scandalously in arrear, and I shall get into terrible hot water if I do not clear off some things that have been hanging about me for months and years.

If you will send me up the specimens, however, I will ask Flower (whom I see constantly) to examine them for you. The examination will be no great trouble, and I am ashamed to make a fuss about it, but I have sworn a big oath to take no fresh work, great or small, until certain things are done.

I wake up in the morning with somebody saying in my ear, "A is not done, and B is not done, and C is not done, and D is not done," etc., and a feeling like a fellow whose duns are all in the street waiting for him. By the way, you ask me what I am doing now, so I will just enumerate some of the A, B, and C's aforesaid.

A. Editing lectures on Vertebrate skull and bringing them out in the *Medical Times*.

B. Editing and re-writing lectures on Elementary Physiology,[1] just delivered here and reported as I went along.

C. Thinking of my course of twenty-four lectures on the Mammalia at Coll. Surgeons in next spring, and making investigations bearing on the same.

D. Thinking of and working at a *Manual of Compara-*

---

[1] Delivered on Friday evenings from April to June at Jermyn Street, and reported in the *Medical Times*. They formed the basis of his well-known little book on *Elementary Physiology*, published 1866. He writes on April 22 :—" Macmillan has just been with me, and I am let in for a school book on physiology based on these lectures of mine. Money arrangements not quite fixed yet, but he is a good fellow, and will not do me unnecessarily."

tive Anatomy (may it be d—d), which I have had in hand these seven years.

E. Getting heaps of remains of new Labyrinthodonts from the Glasgow coalfield, which have to be described.

F. Working at a memoir on *Glyptodon* based on a new and almost entire specimen at the College of Surgeons.

G. Preparing a new decade upon Fossil fishes for this place.

H. Knowing that I ought to have written long ago a description of a most interesting lot of Indian fossils sent to me by Oldham.

I. Being blown up by Hooker for doing nothing for the *Natural History Review.*

K. Being bothered by sundry editors just to write articles "which you know you can knock off in a moment."

L. Consciousness of having left unwritten letters which ought to have been written long ago, especially to C. Darwin.

M. General worry and botheration. Ten or twelve people taking up my time all day about their own affairs.

N. O. P. Q. R. S. T. U. V. W. X. Y. Z.

Societies.

Clubs.

Dinners, evening parties, and all the apparatus for wasting time called "Society." Colensoism and botheration about Moses. . . . Finally pestered to death in public and private because I am supposed to be what they call a "Darwinian."

If that is not enough, I could exhaust the Greek alphabet for heads in addition.

I am glad to hear that Wyman thinks well of my book, as he is very competent to judge. I hear it is republished in America, but I suppose I shall get nothing out of it.

An undated letter to Kingsley, who had suggested

that he should write an article on Prayer, belongs probably to the autumn of 1863 :—

I should like very much to write such an article as you suggest, but I am very doubtful about undertaking it for *Fraser*. Anything I could say would go to the root of praying altogether, for inasmuch as the whole universe is governed, so far as I can see, in the same way, and the moral world is as much governed by laws as the physical —whatever militates against asking for one sort of blessing seems to me to tell with the same force against asking for any other.

Not that I mean for a moment to say that prayer is illogical, for if the whole universe is ruled by fixed laws it is just as logically absurd for me to ask you to answer this letter as to ask the Almighty to alter the weather. The whole argument is an "old foe with a new face," the freedom and necessity question over again.

If I were to write about the question I should have to develop all this side of the problem, and then having shown that logic, as always happens when it is carried to extremes, leaves us *bombinantes in vacuo*, I should appeal to experience to show that prayers of this sort are not answered, and to science to prove that if they were they would do a great deal of harm.

But you know this would never do for the atmosphere of *Fraser*. It would be much better suited for an article in my favourite organ, the wicked *Westminster*.

However, to say truth, I do not see how I am to undertake anything fresh just at present. I have promised an article for *Macmillan* ages ago; and Masson scowls at me whenever we meet. I am afraid to go through the Albany lest Cook [1] should demand certain reviews of books which have been long in my hands. I

---

[1] John Douglas Cook, 1808-1868, editor of the *Saturday Review* from 1858.

am just completing a long memoir for the Linnean Society; a monograph on certain fossil reptiles must be finished before the new year. My lectures have begun, and there is a certain "Manual" looming in the background. And to crown all, these late events [1] have given me such a wrench that I feel I must be prudent.

The following reference to Robert Lowe, afterwards Lord Sherbrooke, has a quasi-prophetic interest:—

May 7.—Dined at the Smiths' [2] last night. Lowe was to have been there, but had a dinner-party of his own. . . . I have come to the conviction that our friend Bob is a most admirable, well-judging statesman, for he says I am the only man fit to be at the head of the British Museum,[3] and that if he had his way he would put me there.

Years afterwards, on Sir R. Owen's retirement, he was offered the post, but declined it, as he greatly disliked the kind of work. At the same time, he pointed out to the Minister who made the offer that the man of all others for the post would be the late distinguished holder of it, Sir W. H. Flower, a suggestion happily acted on.

Early in August a severe loss befell him in the sudden death of his brother George, who had been his close friend ever since he had returned from Australia, and had given him all the help and sympathy in his struggles that could be given by a man of the world without special interests in science

---

[1] The death of his brother.
[2] Dr. (afterwards Sir William) Smith, of Dictionary fame.
[3] *I.e.* of the Natural History Collections.

or literature. With brilliancy enough to have won
success if he had had patience to ensure it, he was not
only a pleasant companion, a "clubbable man" in
Johnson's phrase, but a friend to trust. The two
households had seen much of one another; the child-
less couple regarded their brother's children almost
as their own. Thus a real gap was made in the
family circle, and the trouble was not lessened by the
fact that George Huxley's affairs were left in great
confusion, and his brother not only spent a great
deal of time in looking after the interests of the
widow, but took upon himself certain obligations in
order to make things straight, with the result that he
was even compelled to part with his Royal Medal,
the gold of which was worth £50.

CHAPTER XVIII

1864

THE year 1864 was much like 1863. The Hunterian Lectures were still part of his regular work. The Fishery Commission claimed a large portion of his time. From March 28 to April 2 he was in Cornwall; on May 7 at Shoreham; from July 24 to September 9 visiting the coasts of Scotland and Ireland. The same pressure of work continued. He published four papers on paleontological or anatomical subjects in the *Natural History Review*,[1] he wrote "Further Remarks upon the Human Remains from the Neanderthal," and later (see pp. 365 and 385), dealing with "Criticisms of the *Origin of Species*" (*Collected Essays*, ii. 80, "Darwiniana"), he gently but firmly dispersed several misconceptions of his old friend Kölliker as to the plain meaning of the

[1] On "Cetacean Fossils termed Ziphius by Cuvier," in the *Transactions of the Geological Society*; in those of the *Zoological*, papers on "Arctocebus Calabarensis" and "the Structure of the Stomach in Desmodus Rufus"; and on the "Osteology of the Genus Glyptodon," in the *Phil. Trans.*

book; and ridiculed the pretentious ignorance of M. Flourens' dicta upon the same subject; while in the winter he delivered a course of lectures to working men on "The Various Races of Mankind," a choice of subject which shows that his chief interest at that time lay in Ethnology.

JERMYN STREET, *Jan.* 16, 1864.

MY DEAR DARWIN—I have had no news of you for a long time, but I earnestly hope you are better.

Have you any objection to putting your name to Flower's certificate for the Royal Society herewith enclosed? It will please him much if you will; and I go bail for his being a thoroughly good man in all senses of the word —which, as you know, is more than I would say for everybody.

Don't write any reply ; but Mrs. Darwin perhaps will do me the kindness to send the thing on to Lyell as per enclosed envelope. I will write him a note about it.

We are all well, barring customary colds and various forms of infantile pip. As for myself, I am flourishing like a green bay tree (appropriate comparison, Soapy Sam would observe), in consequence of having utterly renounced societies and society since October.

I have been working like a horse, however, and shall work "horser" as my college lectures begin in February.
—*Tout à vous,* T. H. HUXLEY.

ROYAL SCHOOL OF MINES,
JERMYN STREET, *April* 18, 1864.

MY DEAR DARWIN—I was rejoiced to see your hand-writing again, so much so that I shall not scold you for undertaking the needless exertion (as it's my duty to do) of writing to thank me for my book.[1]

---

[1] *Hunterian Lectures on Anatomy.*

I thought the last lecture would be nuts for you, but it is really shocking. There is not the smallest question that Owen wrote both the article " Oken " and the *Archetype* Book, which appeared in its second edition in French —why, I know not. I think that if you will look at what I say again, there will not be much doubt left in your mind as to the identity of the writer of the two.

The news you give of yourself is most encouraging ; but pray don't think of doing any work again yet. Careful as I have been during this last winter not to burn the candle at both ends, I have found myself, since the pressure of my lectures ceased, in considerable need of quiet, and I have been lazy accordingly.

I don't know that I fear, with you, caring too much for science—for there are lots of other things I should like to go into as well, but I do lament more and more as time goes on, the necessity of becoming more and more absorbed in one kind of work, a necessity which is created for any one in my position, partly by one's reputation, and partly by one's children. For directly a man gets the smallest repute in any branch of science, the world immediately credits him with knowing about ten times as much as he really does, and he becomes bound in common honesty to do his best to climb up to his reputed place. And then the babies are a devouring fire, eating up the present and discounting the future ; they are sure to want all the money one can earn, and to be the better for all the credit one can win.

However, I should fare badly without the young monkeys. Your pet Marian is almost as shy as ever, though she has left off saying " can't," by the way.

My wife is wonderfully well. As I tell her, Providence has appointed her to take care of me when I am broken down and decrepit.

I hope you can say as much of Mrs. Darwin. Pray give her my kind regards.—And believe me, ever yours faithfully,   T. H. HUXLEY.

A letter to his sister gives a sketch of his position
at this time, speaking of which he says to Dr.
(afterwards Sir J.) Fayrer, "You and I have travelled
a long way, in all senses, since you settled my career
for me on the steps of the Charing Cross Hospital."[1]
It must be remembered that his sister was living in
Tennessee, and that her son at fifteen was serving in
the Confederate army.

JERMYN STREET, 4/5/64.

You will want to know something about my progress
in the world. Well, at this moment I am Professor of
Natural History here, and Hunterian Professor of
Comparative Anatomy at the College of Surgeons. The
former is the appointment I have held since 1855 ; the
latter chair I was asked to take last year, and now I
have delivered two courses in that famous black gown
with the red facings which the doctor will recollect very
well. What with the duties of these two posts and other
official and non-official business, I am worked to the full
stretch of my powers, and sometimes a little beyond
them ; though hitherto I have stood the wear and tear
very well.

I believe I have won myself a pretty fair place in
science, but in addition to that I have the reputation (of
which, I fear, you will not approve) of being a great
heretic and a savage controversialist always in rows. To
the accusation of heresy I fear I must plead guilty ; but
the second charge proceeds only, I do assure you, from a
certain unconquerable hatred of lies and humbug which
I cannot get over.

I have read all you tell me about the south with
much interest and with the warmest sympathy, so far as

---

[1] Sir Joseph Fayrer went in 1850 to India. He became a
professor at the Medical College, Calcutta, and afterwards President of the Medical Board, India Office.

the fate of the south affects you. But I am in the condition of most thoughtful Englishmen. My heart goes with the south, and my head with the north.

I have no love for the Yankees, and I delight in the energy and self-sacrifice of your people ; but for all that, I cannot doubt that whether you beat the Yankees or not, you are struggling to uphold a system which must, sooner or later, break down.

I have not the smallest sentimental sympathy with the negro ; don't believe in him at all, in short. But it is clear to me that slavery means, for the white man, bad political economy ; bad social morality ; bad internal political organisation, and a bad influence upon free labour and freedom all over the world. For the sake of the white man, therefore, for your children and grand-children, directly, and for mine, indirectly, I wish to see this system ended.[1] Would that the south had had the wisdom to initiate that end without this miserable war !

All this must jar upon you sadly, and I grieve that it does so ; but I could not pretend to be other than I am, even to please you. Let us agree to differ upon this point. If I were in your place I doubt not I should feel as you do ; and, when I think of you, I put myself in your place and feel with you as your brother Tom. The learned gentleman who has public opinions for which he is responsible is another "party" who walks about in T's clothes when he is not thinking of his sister.

If this were not my birthday I should not feel justified in taking a morning's holiday to write this long letter to you. The ghosts of undone pieces of work are dancing about me, and I must come to an end.

---

[1] Cf. *Reader*, February 27 onwards, where these general arguments against slavery appear in a controversy arising from his ninth Hunterian Lecture, in which, while admitting negro inferiority, he refutes those who justify slavery on the ground that physiologically the negro is very low in the scale.

Give my love to your husband. I am glad to hear he wears so well. And don't forget to give your children kindly thoughts of their uncle. Dr. Wright gives a great account of my namesake, and says he is the handsomest youngster in the Southern States. That comes of his being named after me, you know how renowned for personal beauty I always was.

I asked Dr. Wright if you had taken to spectacles, and he seemed to think not. I had a pain about my eyes a few months ago, but I found spectacles made this rather worse and left them off again. However, I do catch myself holding a newspaper further off than I used to do.

Now don't let six months go by without writing again. If our little venture succeeds this time, we shall send again.[1]—Ever, my dearest Lizzie, your affectionate brother,   T. H. Huxley

He writes to his wife, who had taken the children to Margate :—

Sept. 22.—I am now busy over a paper for the Zool. Soc. ; after that there is one for the Ethnological which was read last session though not written. . . . Don't blaspheme about going into the bye-ways. They are both in the direct road of the book, only over the hills instead of going over the beaten path.

Oct. 6.—I heard from Darwin last night jubilating over an article of mine which is published in the last number of the *Nat. Hist. Review*, and which he is immensely pleased with. . . . My lectures tire me, from want of practice, I suppose. I shall soon get into swing.

The article in question was the "Criticisms of the *Origin of Species*," of which he writes to Darwin :—

---

[1] *I.e.* a package of various presents to the family.

JERMYN STREET, *Oct.* 5, 1864.

MY DEAR DARWIN—I am very glad to see your hand-writing (in ink) again, and none the less on account of the pretty words into which it was shaped.

It is a great pleasure to me that you like the article, for it was written very hurriedly, and I did not feel sure when I had done that I had always rightly represented your views.

Hang the two scalps up in your wigwam !

Flourens I could have believed anything of, but how a man of Kölliker's real intelligence and ability could have so misunderstood the question is more than I can comprehend.

It will be a thousand pities, however, if any review interferes with your saying something on the subject yourself. Unless it should give you needless work I heartily wish you would.

Everybody tells me I am looking so exceedingly well that I am ashamed to say a word to the contrary. But the fact is, I get no exercise, and a great deal of bothering work on our Commission's Cruise ; and though much fatter (indeed a regular bloater myself), I am not up to the mark. Next year I will have a real holiday.[1]

I am a bachelor, my wife and belongings being all at that beautiful place, Margate. When I came back I found them all looking so seedy that I took them off bag and baggage to that, as the handiest place, before a week was over. They are wonderfully improved already, my wife especially being abundantly provided with her favourite east wind. Your godson is growing a very sturdy fellow, and I begin to puzzle my head with thinking what he is and what he is not to be taught.

Please to remember me very kindly to Mrs. Darwin, and believe me yours very faithfully, T. H. HUXLEY.

---

[1] At the end of the year, as so often, he went off for a ploy with Tyndall, this time into Derbyshire, walking vigorously over the moors.

The following illustrates the value he set upon public examinations as a practical means for spreading scientific education, and upon first-rate examiners as a safeguard of proper methods of teaching.

Oct. 6, 1864.

MY DEAR HOOKER—Donnelly told me to-day that you had been applied to by the Science and Tarts Department to examine for them in botany, and that you had declined.

Will you reconsider the matter ?   I have always taken a very great interest in the science examinations, looking upon them, as I do, as the most important engine for forcing science into ordinary education.

The English nation will not take science from above, so it must get it from below.

Having known these examinations from the beginning, I can assure you that they are very genuine things, and are working excellently.   And what I have regretted from the first is that the botanical business was not taken in hand by you, instead of by ——.

Now, like a good fellow, think better of it.   The papers are necessarily very simple, and one of Oliver's pupils could look them over for you.   Let us have your co-operation and the advantage of that reputation for honesty and earnestness which you have contrived (Heaven knows how) to get.

I have come back fat and seedy for want of exercise. All my belongings are at Margate.   Hope you don't think my review of Darwin's critics too heretical if you have seen it.—Ever yours faithfully,   T. H. HUXLEY.

When is our plan for getting some kind of meetings during the winter to be organised ?

The next two letters refer to the award of the Copley Medal to Mr. Darwin.   Huxley was exceedingly indignant at an attempt on the part of the president to discredit the *Origin* by a side wind :—

JERMYN STREET, *Nov.* 4, 1864.

MY DEAR DARWIN—I write two lines which are *not to be answered*, just to say how delighted 1 am at the result of the doings of the Council of the Royal Society yesterday. Many of us were somewhat doubtful of the result, and the more ferocious sort had begun to whet their beaks and sharpen their claws in preparation for taking a very decided course of action had there been any failure of justice this time. But the affair was settled by a splendid majority, and our ruffled feathers are smoothed down.

Your well-won reputation would not have been lessened by the lack of the Copley, but it would have been an indelible reproach to the Royal Society not to have given it you, and a good many of us had no notion of being made to share that ignominy.

But quite apart from all these grand public-spirited motives and their results, you ought as a philanthropist to be rejoiced in the great satisfaction the award has given to your troops of friends, to none more than my wife (whom I woke up to tell the news when I got home late last night).—Yours ever,　　T. H. HUXLEY.

Please remember us kindly to Mrs. Darwin, and make our congratulations to her on owning a Copley medallist.

JERMYN STREET, *Dec.* 3, 1864.

MY DEAR HOOKER—I wish you had been at the Anniversary Meeting and Dinner, because the latter was very pleasant, and the former, to me, very disagreeable. My distrust of Sabine is as you know chronic, and I went determined to keep careful watch on his address, lest some crafty phrase injurious to Darwin should be introduced. My suspicions were justified, the only part of the address to Darwin written by Sabine himself containing the following passage :—

"Speaking generally and collectively, we have expressly omitted it (Darwin's theory) from the grounds of our award."

Of course this would be interpreted by everybody as meaning that, after due discussion, the council had formally resolved not only to exclude Darwin's theory from the grounds of the award, but to give public notice through the president that they had done so, and furthermore, that Darwin's friends had been base enough to accept an honour for him on the understanding that in receiving it he should be publicly insulted !

I felt that this would never do, and therefore when the resolution for printing the address was moved, I made a speech which I took care to keep perfectly cool and temperate, disavowing all intention of interfering with the liberty of the president to say what he pleased, but exercising my constitutional right of requiring the minutes of council making the award to be read, in order that the Society might be informed whether the conditions implied by Sabine had been imposed or not.

The resolution was read, and of course nothing of the kind appeared. Sabine didn't exactly like it, I believe. Both Busk and Falconer remonstrated against the passage to him, and I hope it will be withdrawn when the address is printed.[1]

If not, there will be an awful row, and I for one will show no mercy.—Ever yours faithfully,

T. H. HUXLEY.

The foundation of the *x* Club towards the end of 1864 was a notable event for Huxley and his circle of scientific friends. It was growing more and more difficult for them to see one another except now and again at meetings of the learned societies, and even

---

[1] The passage stands in the published address, but followed by another passage which softens it down.

that was quite uncertain. The pressure of Huxley's own work may be inferred from his letters at this time (especially to Darwin, July 2, 1863, and January 16, 1864). Not only society, but societies had to be almost entirely given up. Moreover, the distance from one another at which some of these friends lived added another difficulty, so that Huxley writes to Hooker in his "remote province" of Kew: "I wonder if we are ever to meet again in this world." Accordingly in January 1864, Hooker gladly embraced a proposal of Huxley's to organise some kind of regular meeting, a proposal which bore fruit in the establishment of the *x* Club. On November 3, 1864, the first meeting was held at St. George's Hotel, Albemarle Street, where they resolved to dine regularly "except when Benham cannot have us, in which case dine at the Athenæum." In the latter eighties, however, the Athenæum became the regular place of meeting, and it was here that the "coming of age" of the club was celebrated in 1885.

Eight members met at the first meeting; the second meeting brought their numbers up to nine by the addition of W. Spottiswoode, but the proposal to elect a tenth member was never carried out. On the principle of *lucus a non lucendo*, this lent an additional appropriateness to the symbol *x*, the origin of which Huxley thus describes in his reminiscences of Tyndall in the *Nineteenth Century* for January 1894 :—

At starting, our minds were terribly exercised over the name and constitution of our society. As opinions on this grave matter were no less numerous than the members—indeed more so—we finally accepted the happy suggestion of our mathematicians to call it the x Club ; and the proposal of some genius among us, that we should have no rules, save the unwritten law not to have any, was carried by acclamation.

Besides Huxley, the members of the club were as follows :—

George Busk, F.R.S. (1807-87), then secretary of the Linnean Society, a skilful anatomist.[1]

Edward Frankland (1825-1899), For. Sec. R.S., K.C.B., then Professor of Chemistry in the Royal Institution, and afterwards at the Royal College of Science.

Thomas Archer Hirst, F.R.S., then mathematical master at University College School.[2]

Joseph Dalton Hooker, F.R.S., K.C.S.I., Pres. R.S. 1873, the great botanist, then Assistant Director at Kew Gardens to his father, Sir William Hooker.

Sir John Lubbock, Bart., F.R.S., M.P., now Lord Avebury, the youngest of the nine, who had already made his mark in archæology, and was then preparing to bring out his *Prehistoric Times*.

[1] He served as surgeon to the hospital ship *Dreadnought* at Greenwich till 1856, when he resigned, and, retiring from practice, devoted himself to scientific pursuits, and was elected President of the College of Surgeons in 1871.

[2] In 1865 appointed Professor of Physics ; in 1867, of Pure Mathematics, at University College, London ; and from 1873 to 1883 Director of Naval Studies at the Royal Naval College, Greenwich ; an old Marburg student, and intimate friend of Tyndall, whom he had succeeded at Queenwood College in 1853. He died in 1892.

Herbert Spencer, who had already published *Social Statics*, *Principles of Psychology*, and *First Principles*.

William Spottiswoode (1825-1883), F.R.S., Treasurer and afterwards President R.S. 1878, who carried on the business of the Queen's printer as well as being deeply versed in mathematics, philosophy, and languages.

John Tyndall, F.R.S. (1820-1893), who had been for the last eleven years Professor of Natural Philosophy at the Royal Institution, where he succeeded Faraday as superintendent.

The one object, then, of the club was to afford a certain meeting-ground for a few friends who were bound together by personal regard and community of scientific interests, yet were in danger of drifting apart under the stress of circumstances. They dined together on the first Thursday in each month, except July, August, and September, before the meeting of the Royal Society, of which all were members excepting Mr. Spencer, the usual dining hour being six, so that they should be in good time for the society's meeting at eight; and a minute of December 5, 1885, when Huxley was treasurer and revived the ancient custom of making some note of the conversation, throws light on the habits of the club. "Got scolded," he writes, "for dining at 6.30. Had to prove we have dined at 6.30 for a long time by evidence of waiter." (At the February meeting, however, "agreed to fix dinner hour six hereafter.") "Talked politics, scandal, and the three classes of witnesses—liars,

d—d liars, and experts.   Huxley gave account of
civil list pension.   Sat to the unexampled hour of
10 P.M., except Lubbock, who had to go to
Linnean."

For some time there was a summer meeting which
consisted of a week-end excursion of members and
their wives (x's+yv's, as the correct formula ran) to
some place like Burnham or Maidenhead, Oxford or
Windsor; but this grew increasingly difficult to
arrange, and dropped before very long.

Guests were not excluded from the dinners of the
club; men of science or letters of almost every
nationality dined with the x at one time or another;
Darwin, W. K. Clifford, Colenso, Strachey, Tolle-
mache, Helps; Professors Bain, Masson, Robertson
Smith, and Bentham the botanist, Mr. John Morley,
Sir D. Galton, Mr. Jodrell, the founder of several
scientific lectureships; Dr. Klein; the Americans
Marsh, Gilman, A. Agassiz, and Youmans, the latter
of whom met here several of the contributors to the
International Science Series organised by him; and
continental representatives, as Helmholtz, Laugel,
and Cornu.

Small as the club was, the members of it were
destined to play a considerable part in the history of
English science.   Five of them received the Royal
Medal; three the Copley; one the Rumford; six
were Presidents of the British Association; three
Associates of the Institute of France; and from
amongst them the Royal Society chose a Secretary, a

Foreign Secretary, a Treasurer, and three successive Presidents.

I think, originally (writes Huxley, *l.c.*) there was some vague notion of associating representatives of each branch of science; at any rate, the nine who eventually came together could have managed, among us, to contribute most of the articles to a scientific Encyclopædia.

They included leading representatives of half a dozen branches of science :—mathematics, physics, philosophy, chemistry, botany, and biology; and all were animated by similar ideas of the high function of science, and of the great Society which should be the chief representative of science in this country. However unnecessary, it was perhaps not unnatural that a certain jealousy of the club and its possible influence grew up in some quarters. But whatever influence fell to it as it were incidentally—and earnest men with such opportunities of mutual understanding and such ideals of action could not fail to have some influence on the progress of scientific organisation— it was assuredly not sectarian nor exerted for party purposes during the twenty-eight years of the club's existence.

I believe that the *x* (continues Huxley) had the credit of being a sort of scientific caucus, or ring, with some people. In fact, two distinguished scientific colleagues of mine once carried on a conversation (which I gravely ignored) across me, in the smoking-room of the Athenæum, to this effect, " I say, A., do you know anything about the *x* Club ? " "Oh yes, B., I have heard of it. What do

they do?"   "Well, they govern scientific affairs, and
really, on the whole, they don't do it badly."   If my good
friends could only have been present at a few of our
meetings, they would have formed a much less exalted
idea of us, and would, I fear, have been much shocked
at the sadly frivolous tone of our ordinary conversation.

The *x* Club is probably unique in the smallness of
its numbers, the intellectual eminence of its members,
and the length of its unchanged existence.   The
nearest parallel is to be found in "The Club."[1]   Like
the *x*, "The Club" began with eight members at its
first meeting, and of the original members Johnson
lived twenty years, Reynolds twenty-eight, Burke
thirty-three, and Bennet Langton thirty-seven.   But
the ranks were earlier broken.   Within ten years
Goldsmith died, and he was followed in a twelvemonth
by Nugent, and five years later by Beauclerk and
Chamier.   Moreover, the eight were soon increased
to twelve ; then to twenty and finally to forty, while
the gaps were filled up as they occurred.

In the *x*, on the contrary, nearly nineteen years
passed before the original circle was broken by the
death of Spottiswoode.   From 1864 to Spottiswoode's
death in 1883 the original circle remained unbroken ;
the meetings "were steadily continued for some
twenty years, before our ranks began to thin; and
one by one, *geistige Naturen* such as those for which

---

[1] Of which Huxley was elected a member in 1884.   Tyndall
and Hooker were also members.

the poet [1] so willingly paid the ferryman, silent but
not unregarded, took the vacated places." The peculiar
constitution of the club scarcely seemed to admit of
new members; not, at all events, without altering
the unique relation of friendship joined to common
experience of struggle and success which had lasted
so long. After the death of Spottiswoode and Busk,
and the ill-health of other members, the election of
new members was indeed mooted, but the proposal
was ultimately negatived. Huxley's opinion on this
point appears from letters to Sir E. Frankland in
1886 and to Sir J. D. Hooker in 1888.

As for the filling up the vacancies in the *x*, I am disposed to take Tyndall's view of the matter. Our little
club had no very definite object beyond preventing a few
men who were united by strong personal sympathies from
drifting apart by the pressure of busy lives.

Nobody could have foreseen or expected twenty odd
years ago when we first met, that we were destined to
play the parts we have since played, and it is in the nature
of things impossible that any of the new members proposed
(much as we may like and respect them all), can carry on
the work which has so strangely fallen to us.

An axe with a new head and a new handle may be the
same axe in one sense, but it is not the familiar friend
with which one has cut one's way through wood and brier.

And in the other letter—

What with the lame dog condition of Tyndall and

---

[1] Nimm dann Führmann,.
Nimm die Miethe
Die Ich gerne dreifach biete;
Zwei, die eben überfuhren
Waren geistige Naturen.

Hirst and Spencer and my own recurrent illnesses, the x is not satisfactory. But I don't see that much will come from putting new patches in. The x really has no *raison d'être* beyond the personal attachment of its original members. Frankland told me of the names that had been mentioned, and none could be more personally welcome to me . . . but somehow or other they seem out of place in the x.

However, I am not going to stand out against the general wish, and I shall agree to anything that is desired.

Again—

The club has never had any purpose except the purely personal object of bringing together a few friends who did not want to drift apart. It has happened that these cronies had developed into bigwigs of various kinds, and therefore the club has incidentally—I might say accidentally—had a good deal of influence in the scientific world. But if I had to propose to a man to join, and he were to say, well, what is your object? I should have to reply like the needy knife-grinder, "Object, God bless you, sir, we've none to show."

As he wrote elsewhere (*Nineteenth Century*, Jan. 1894 ; see p. 369 above)—

Later on, there were attempts to add other members, which at last became wearisome, and had to be arrested by the agreement that no proposition of that kind should be entertained, unless the name of the new member suggested contained all the consonants absent from the names of the old ones. In the lack of Slavonic friends this decision put an end to the possibility of increase.

After the death, in February 1892, of Hirst, a most devoted supporter of the club, who " would, I believe, represent it in his sole person rather than pass the day over," only one more meeting took place, in the following month. With five of the six survivors domiciled far from town, meeting after meeting fell through, until the treasurer wrote, "My idea is that it is best to let it die out unobserved, and say nothing about its decease to any one."

Thus it came to pass that the March meeting of the club in 1892 remained its last. No ceremony ushered it out of existence. Its end exemplified a saying of Sir J. Hooker's, "At our ages clubs are an anachronism." It had met 240 times, yet, curious to say, although the average attendance up to 1883 was seven out of nine, the full strength of the club only met on twenty-seven occasions.

CHAPTER XIX

1865

THE progress of the American civil war suggested to Huxley in 1865 the text for an article "Emancipation, Black and White," the emancipation of the negro in America and the emancipation of women in England, which appeared in the *Reader* of May 20 (*Coll. Ess.* iii. 66). His main argument for the emancipation of the negro was that already given in his letter to his sister (p. 362); namely, that in accordance with the moral law that no human being can arbitrarily dominate over another without grievous damage to his own nature, the master will benefit by freedom more than the freed-man. And just as the negro will never take the highest places in civilisation yet need not to be confined to the lowest, so, he argues, it will be with women. "Nature's old salique law will never be repealed, and no change of dynasty will be effected," although "whatever argument justifies a given education for all boys justifies its application to girls as well."

With this may be compared his letter to the *Times* of July 8, 1874 (ii. pp. 139, 140).

No scientific monographs were published in 1865 by Huxley, but his lectures of the previous winter to working men on "The Various Races of Mankind" are an indication of his continued interest in Ethnology, which, set going, as has been said, by the promise to revise the woodcuts for Lyell's book, found expression in such papers as the "Human Remains in the Shell Mounds," 1863; the "Neanderthal Remains" of 1864; the "Methods and Results of Ethnology" of 1865; his Fullerian Lectures of 1866-67; papers on "Two Widely Contrasted Forms of the Human Cranium" of 1866 and 1868; the "Patagonian Skulls" of 1868; and "Some Fixed Points in British Ethnology" of 1870.

His published ethnological papers (says Sir Michael Foster) are not numerous, nor can they be taken as a measure of his influence on this branch of study. In many ways he has made himself felt, not the least by the severity with which on the one hand he repressed the pretensions of shallow persons who, taking advantage of the glamour of the Darwinian doctrine, talked nonsense in the name of anthropological science, and on the other hand, exposed those who in the structure of the brain or of other parts, saw an impassable gulf between man and the monkey. The episode of the "hippocampus" stirred for a while not only science but the general public. He used his influence, already year by year growing more and more powerful, to keep the study of the natural history of man within its proper lines, and chiefly with

this end in view held the Presidential Chair of the
Ethnological Society in 1869-70.   It was mainly through
his influence that this older Ethnological Society was, a
year later, in 1871, amalgamated with a newer rival
society, the Anthropological, under the title of "The
Anthropological Institute."

One of the leading points in the "Methods and
Results of Ethnology," repeatedly insisted on in
later years (see ii. 7, 250, and iii. 172) is the
rejection of the prevalent theory that community
of language implies community of race, Shleicher
indeed asserting that language is a more constant
character than cranial peculiarities.   We talk of
Latin and Teutonic races, of Celtic and English,
dividing them in fact by their language, with no
more accuracy than if an archæologist of the future
were to call the negroes of Hayti a European
stock because they spoke French.   "Community
of language testifies to close contact between the
people who speak the language, but to nothing else."

In this connection he writes to Prof. Max Müller
on June 15, 1865 :—

My DEAR SIR—I beg your acceptance of the number
of the *Fortnightly Review* containing my article on
Ethnology, which accompanies this note.

I lost no time, on Monday, in referring to
"Christianity and Mankind" ; and the perusal of your
chapter on Ethnology *v.* Phonology led me profoundly
to regret that I had not been able to avail myself of the
aid of so powerful an ally.

But if you will continue to pull one way and I the
other, I have hopes we shall be able to get Ethnology and

Phonology apart in time.—Ever, my dear Sir, very faithfully yours,   T. H. HUXLEY.

During this time he was constantly occupied with paleontological work, as the following letter to Sir C. Lyell indicates—

JERMYN STREET, *Nov.* 27, 1865.

MY DEAR SIR CHARLES—I returned last night from a hasty journey to Ireland, whither I betook myself on Thursday night, being attracted vulture-wise by the scent of a quantity of carboniferous corpses. The journey was as well worth the trouble as any I ever undertook, seeing that in a morning's work I turned out ten genera of vertebrate animals of which five are certainly new ; and of these four are Labyrinthodonts, amphibia of new types. These four are baptized *Ophiderpeton, Lepterpeton, Ichthyerpeton, Keraterpeton.* They all have ossified spinal columns and limbs. The special interest attaching to the two first is that they represent a type of Labyrinthodonts hitherto unknown, and corresponding with *Siren* and *Amphiuma* among living Amphibia. Ophiderpeton, for example, is like an eel, about three feet long with small fore legs and rudimentary hind ones.

In the year of grace 1861, there were three genera of European carboniferous Labyrinthodonts known, *Archegosaurus, Scleroceplus, Parabatrachus.*

The vertebral column of Archegosaurus was alone known, and it was in a remarkably imperfect state of ossification. Since that dâte, by a succession of odd chances, seven new genera have come into my hands, and of these six certainly have well-ossified and developed vertebral columns.

I reckon there are now about thirty genera of Labyrinthodonts known from all parts of the world and all deposits. Of these eleven have been established by myself in the course of the last half-dozen years, upon remains which have come into my hands by the merest chance.

Five and twenty years ago, all the world but yourself believed that a vertebrate animal of higher organisation than a fish in the carboniferous rocks never existed. I think the whole story is not a bad comment upon negative evidence.

Jan. 1, 1865.

MY DEAR DARWIN—I cannot do better than write my first letter of the year to you, if it is only to wish you and yours your fair share (and more than your fair share, if need be) of good for the New Year. The immediate cause of my writing, however, was turning out my pocket and finding therein an unanswered letter of yours containing a scrap on which is a request for a photograph, which I am afraid I overlooked. At least I hope I did, and then my manners won't be so bad. I enclose the latest version of myself.

I wish I could follow out your suggestion about a book on zoology. (By the way, please to tell Miss Emma that my last book *is* a book.[1] Marry come up! Does her ladyship call it a pamphlet?)

But I assure you that writing is a perfect pest to me unless I am interested, and not only a bore but a very slow process. I have some popular lectures on Physiology,[2] which have been half done for more than a twelvemonth, and I hate the sight of them because the subject no longer interests me, and my head is full of other matters.

---

[1] The first volume of his Hunterian Lectures on *Comparative Anatomy*. A second volume never appeared. Miss Darwin, as her father wrote to Huxley after the delivery of his Working Men's Lectures in 1862, "was reading your Lectures, and ended by saying, 'I wish he would write a book.' I answered, 'he has just written a great book on the skull.' 'I don't call that a book,' she replied, and added, 'I want something that people can read; he does write so well.'"

[2] See letter of April 22, 1863.

So I have just done giving a set of lectures to working men on "The Various Races of Mankind," which really would make a book in Miss Emma's sense of the word, and which I have had reported. But when am I to work them up? Twenty-four Hunterian Lectures loom between me and Easter, I am dying to get out the second volume of the book that is not a book, but in vain.

I trust you are better, though the last news I had of you from Lubbock was not so encouraging as I could have wished.

With best wishes and remembrances to Mrs. Darwin. —Ever yours,   T. H. HUXLEY.

Thanks for "Für Darwin," [1] I had it.

26 ABBEY PLACE, *Jan.* 15, 1865.

MY DEAR DARWIN—Many thanks for Deslongchamps paper, which I do not possess.

I received another important publication yesterday morning in the shape of a small but hearty son, who came to light a little before six. The wife is getting on capitally, and we are both greatly rejoiced at having another boy, as your godson ran great risks of being spoiled by a harem of sisters.

The leader in the *Reader is* mine, and I am glad you like it. The more so as it has got me into trouble with some of my friends. However, the revolution that is going on is not to be made with rose-water.

I wish if anything occurs to you that would improve the scientific part of the *Reader*, you would let me know, as I am in great measure responsible for it.

I am sorry not to have a better account of your health. With kind remembrances to Mrs. Darwin and the rest of your circle.—Ever yours faithfully,   T. H. HUXLEY.

---

[1] By Fritz Müller, one of Darwin's earliest supporters in Germany.

JERMYN STREET, *May* 1, 1865.

MY DEAR DARWIN—I send you by this post a booklet[1]
none of which is much worth your reading, while of nine-
tenths of it you may say as the man did who had been
trying to read Johnson's *Dictionary*, "that the words were
fine, but he couldn't make much of the story."

But perhaps the young lady who has been kind enough
to act as taster of my books heretofore will read the
explanatory notice, and give me her ideas thereupon
(always recollecting that almost the whole of it was
written in the pre-Darwinian epoch).

I do not hear very good accounts of you—to my
sorrow—though rumours have reached me that the *opus
magnum*[2] is completely developed though not yet born.

I am grinding at the mill and getting a little tired.
My belongings flourishing as I hope you are.—Ever yours
faithfully,   T. H. HUXLEY.

JERMYN STREET, *May* 29, 1865.

MY DEAR DARWIN—I meant to have written to you
yesterday to say how glad I shall be to read whatever
you like to send me.

I have to lecture at the Royal Institution this week,
but after Friday, my time will be more at my own disposal than usual ; and as always I shall be most particularly
glad to be of any use to you.

Any glimmer of light on the question you speak of is
of the utmost importance, and I shall be immensely interested in learning your views.   And of course I need
not add I will do my best to upset them.   That is the
nature of the beast.

I had a letter from one of the ablest of the younger

---

[1] Probably "A Catalogue of the Collection of Fossils in the
Museum of Practical Geology," etc.

[2] *The Variation of Animals and Plants under Domestication,*
published in 1868, one chapter of which is on Pangenesis.

zoologists of Germany, Haeckel, the other day, in which this passage occurs :—

"The Darwinian Theory, the establishment and development of which is the object [of] all my scientific labours, has gained ground immensely in Germany (where it was at first so misunderstood) during the last two years, and I entertain no doubt that it will before long be everywhere victorious." And he adds that I dealt far too mildly with Kölliker.

With kindest remembrances to Mrs. Darwin and your family—Ever yours faithfully,   T. H. Huxley.

This year, as is seen from the foregoing, he was again in direct communication with Professor Ernst Haeckel of Jena, the earliest and strongest champion of Darwinian ideas in Germany. The latter wished to enlarge his observations by joining some English scientific expedition, if any such were in preparation, but was dissuaded by the following reply. The expected book of Darwin's was the *Pangenesis*, and this is also referred to in the three succeeding letters to Darwin himself.

The Royal School of Mines,
Jermyn Street, London, *June 7*, 1865.

My dear Sir—Many thanks for your letter, and for the welcome present of your portrait, which I shall value greatly, and in exchange for which I enclose my own. Indeed I have delayed writing to you in order to be able to send the last "new and improved" edition of myself.

I wish it were in my power to help you to any such appointment as that you wish for. But I do not think our government is likely to send out any scientific expedition to the South Seas. There is a talk about a new Arctic expedition, but I doubt if it will come to

much, and even if it should be organised I could not
recommend your throwing yourself away in an undertaking which promises more frost-bites than anything
else to a naturalist.

In truth, though I have felt and can still feel the
attraction of foreign travel in all its strength, I would
counsel you to stop at home, and as Goethe says, find
your America here. There are plenty of people who
can observe and whose places, if they are expended by
fever or shipwreck, can be well enough filled up. But
there are very few who can grapple with the higher
problems of science as you have done and are doing, and
we cannot afford to lose you. It is the organisation of
knowledge rather than its increase which is wanted just
now. And I think you can help in this great undertaking better in Germany than in New Zealand.

Darwin has been very ill for more than a year past,
so ill, in fact, that his recovery was at one time doubtful.
But he contrives to work in spite of fate, and I hope
that before long we shall have a new book from him.

By way of consolation I sent him an extract from your
letter touching the progress of your views.

I am glad that you did not think my critique of
Kölliker too severe. He is an old friend of mine, and
I desired to be as gentle as possible, while performing
the unpleasant duty of showing how thoroughly he had
misunderstood the question.

I shall look with great interest for your promised book.
Lately I have [been] busy with Ethnological questions,
and I fear I shall not altogether please your able friend
Professor Schleicher in some remarks I have had to make
upon the supposed value of philological evidence.

May we hope to see you at the meeting of the British
Association at Birmingham? It would give many, and
especially myself, much pleasure to become personally
acquainted with you.—Ever yours faithfully,

                                        T. H. HUXLEY.

JERMYN STREET, *June* 1, 1865.

MY DEAR DARWIN—Your MS.[1] reached me safely last evening.

I could not refrain from glancing over it on the spot, and I perceive I shall have to put on my sharpest spectacles and best considering cap.

I shall not write till I have thought well on the whole subject.—Ever yours,   T. H. HUXLEY.

JERMYN STREET, *July* 16, 1865.

MY DEAR DARWIN—I have just counted the pages of your MS. to see that they are all right, and packed it up to send you by post, registered, so I hope it will reach you safely. I should have sent it yesterday, but people came in and bothered me about post time.

I did not at all mean by what I said to stop you from publishing your views, and I really should not like to take that responsibility. Somebody rummaging among your papers half a century hence will find *Pangenesis* and say, "See this wonderful anticipation of our modern theories, and that stupid ass Huxley prevented his publishing them." And then the Carlyleans of that day will make me a text for holding forth upon the difference between mere vulpine sharpness and genius.

I am not going to be made a horrid example of in that way. But all I say is, publish your views, not so much in the shape of formed conclusions, as of hypothetical developments of the only clue at present accessible, and don't give the Philistines more chances of blaspheming than you can help.

I am very grieved to hear that you have been so ill again.—Ever yours faithfully,   T. H. HUXLEY.

26 ABBEY PLACE, *Oct.* 2, 1865.

MY DEAR DARWIN—"This comes hoping you are well," and for no other purpose than to say as much. I

---

[1] The chapter on Pangenesis. See p. 384, note 2.

am just back from seven weeks' idleness at Littlehampton
with my wife and children, the first time I have had a
holiday of any extent with them for years.

We are all flourishing—the babies particularly so—
and I find myself rather loth to begin grinding at the
mill again. There is a vein of laziness in me which crops
out uncommonly strong in your godson, who is about the
idlest, jolliest young four-year-old I know.

You will have been as much grieved as I have been
about dear old Hooker. According to the last accounts,
however, he is mending, and I hope to see him in the
pristine vigour again before long.

My wife has gone to bed or she would join me in the
kindest regards and remembrances to Mrs. Darwin and
your family.—Ever yours faithfully,   T. H. HUXLEY.

The sound judgment and nice sense of honour
for which Huxley was known among his friends often
led those who were in difficulties to appeal to him
for advice. About this time a dispute arose over an
alleged case of unacknowledged "conveyance" of
information. Writing to Hooker, he says the one
party to the quarrel failed to "set the affair straight
with half a dozen words of frank explanation as he
might have done;" as to the other, "like all quiet
and mild men who do get a grievance, he became
about twice as 'wud' as Berserks like you and me."
Both came to him, so that he says, "I have found
it very difficult to deal honestly with both sides
without betraying the confidence of either or making
matters worse." Happily, with his help, matters
reached a peaceful solution, and his final comment
is —

I don't mind fighting to the death in a good big row, but when A and B are supplying themselves from C's orchard, I don't think it is very much worth while to dispute whether B filled his pockets directly from the trees or indirectly helped himself to the contents of A's basket. If B has so helped himself, he certainly ought to say so like a man, but if I were A, I would not much care whether he did or not.

—— has been horribly disgusted about it, but I am not sure the discipline may not have opened his eyes to new and useful aspects of nature.

The summer of 1865 saw the inception of an educational experiment—an International Education Society—to which Huxley gladly gave his support as a step in the right direction. He had long been convinced of the inadequacy of existing forms of education—survivals from the needs of a bygone age—to prepare for the new forms into which intellectual life was passing. That educators should be content to bring up the young generation in the modes of thought which satisfied their forefathers three centuries ago, as if no change had passed over the world since then, filled him with mingled amazement and horror.

The outcome of the scheme was the International College, at Spring Grove, Isleworth, under the headmastership of Dr. Leonhard Schmitz; one of the chief members of the committee being Dr. (afterwards Sir) William Smith, while at the head of the Society was Richard Cobden, under whose presidency it had been registered some time before.

John Stuart Mill, however, refused to join, considering that this was not the most needed reform in
education, and that he could not support a school in
which the ordinary theology was taught.

An article in the *Reader* for June 17, 1865,
sketches the plan. The design was to give a liberal
education to boys whether intended for a profession
or for commerce. The education for both was the
same up to a certain point, corresponding to that
given in our higher schools, together with foreign
languages and the elements of physical and social
science, after which the courses bifurcated.[1] Special
stress was laid on modern languages, both for themselves and as a preparation and help for classical
teaching. Accordingly, the International College
was one of three parallel institutions in England,
France, and Germany, where a boy could in turn
acquire a sound knowledge of all three languages
while continuing the same course of education. The
Franco-Prussian war of 1870, however, proved fatal
to the scheme.

Some letters to his friend Dr. W. K. Parker,[2]
show the good-fellowship which existed between
them, as well as the interest he took in the style and

---

[1] For a fuller account of the scientific education see p. 444.

[2] A man of whom he wrote (preface to Prof. Jeffery Parker's
Life of W. K. Parker, 1893), that "in him the genius of an artist
struggled with that of a philosopher, and not unfrequently the
latter got the worst of the contest." He speaks too of his "minute
accuracy in observation and boundless memory for details and
imagination which absolutely rioted in the scenting out of subtle
and often far-fetched analogies."

success of Parker's work.   Parker was hard at work
on Birds, a subject in which his friend and leader
also was deeply interested, and was indeed preparing
an important book upon it.

Referring to his candidature for the Royal Society,
he writes on February 21, 1865 : "With reference to
your candidature, I am ready to bring your name
forward whenever you like, and to back you with
'all my might, power, amity, and authority,' as
Essex did Bacon (you need not serve me as Bacon
did Essex afterwards), but my impression has been
that you did not wish to come forward this year."

And on November 2, 1866, congratulating him on
his "well-earned honour" of the F.R.S.—"Go on
and prosper.   These are not the things wise men
work for ; but it is not the less proper of a wise man
to take them when they come unsought."

26 ABBEY PLACE, Dec. 3, 1865.

MY DEAR PARKER—I have been so terribly pressed by
my work that I have only just been able to finish the
reading of your paper.

Very few pieces of work which have fallen in my
way come near your account of the Struthious skull in
point of clearness and completeness.   It is a most admirable essay, and will make an epoch in this kind of
inquiry.

I want you, however, to remodel the introduction, and
to make some unessential but convenient difference in the
arrangement of some of the figures.

Secondly, full as the appendix is of most valuable
and interesting matter, I advise you for the present to
keep it back.

My reason is that you have done justice neither to yourself nor to your topics, and that if the appendix is printed as it stands, your labour will be in a great measure lost.

You start subjects enough for half a dozen papers, and partly from the compression thus resulting, and partly from the absence of illustrations, I do not believe there are half a dozen men in Europe who will be able to follow you. Furthermore, though the appendix is relevant enough—every line of it—to those who have dived deep, as you and I have—to any one else it has all the aspects of a string of desultory discussions. *As your father confessor, I forbid the publication of the appendix.* After having had all this trouble with you I am not going to have you waste your powers for want of a little method, so I tell you.

What you are to do is this. You are to rewrite the introduction and to say that the present paper is the first of a series on the structure of the vertebrate skull; that the second will be "On the development of the osseous cranium of the Common Fowl" [and here (if you are good), I will permit you to introduce the episode on cartilage and membrane (illegible)]; the third will be "On the chief modifications of the cranium observed in the Sauropsida."

The fourth, "On the mammalian skull."

The fifth, "On the skull of the Ichthyopsida."

I will give you two years from this time to execute these five memoirs; and then if you have stood good-temperedly the amount of badgering and bullying you will get from me whenever you come dutifully to report progress, you shall be left to your own devices in the third year to publish a paper on "The general structure and theory of the vertebrate skull."

You have a brilliant field before you, and a start such that no one is likely to catch you. Sit deliberately down over against the city, conquer it and make it your own, and don't be wasting powder in knocking down odd bastions with random shells.

I write jestingly, but I really am very much in earnest. Come and have a talk on the matter as soon as you can, for I should send in my report. You will find me in Jermyn Street, Tuesday, Wednesday, or Thursday mornings, Thursday afternoon, but not Tuesday or Wednesday afternoon. Send a line to say when you will come.—Ever yours very faithfully,

T. H. Huxley.

CHAPTER XX

1866

BESIDES his Fullerian lectures on Ethnology at the Royal Institution this year, Huxley published in February 1866 a paper in the *Natural History Review,* on the "Prehistoric Remains of Caithness," based upon a quantity of remains found the previous autumn at Keiss. This, and the article on the "Neanderthal Skull" in the *Natural History Review* for 1864, attracted some notice among foreign anthropologists. Dr. H. Welcker writes about them ; Dr. A. Ecker wants the "Prehistoric Remains" for his new *Archiv für Anthropologie ;* the Société d'Anthropologie de Paris elects him a Foreign Associate.

He was asked by Dr. Fayrer to assist in a great scheme he had proposed to the Asiatic Society,[1] to gather men of every tribe from India, the Malayan Peninsula, Persia, Arabia, the Indian Archipelago, etc., for anthropological purposes. It was well received by the Council of the Society and by the

[1] Comp. Chap. XXIII. *ad init.* and Appendix I.

Lieutenant-Governor of Bengal; anything Huxley could say in its favour would be of great weight. Would he come out as Dr. Fayrer's guest?

Unable to go to Calcutta, he sent the following letter :—

JERMYN STREET, LONDON, *June* 14, 1866.

MY DEAR FAYRER—I lose no time in replying to your second letter, and my first business is to apologise for not having answered the first, but it reached me in the thick of my lectures, and like a great many other things which ought to have been done I put off replying to a more convenient season. I have been terribly hard worked this year, and thought I was going to break down a few weeks ago but luckily I have pulled through.

I heartily wish that there were the smallest chance of my being able to accept your kind invitation and take part in your great scheme at Calcutta. But it is impossible for me to leave England for more than six weeks or two months, and that only in the autumn, a time of year when I imagine Calcutta is not likely to be the scene of anything but cholera patients.

As to your plan itself, I think it a most grand and useful one if it can be properly carried out. But you do things on so grand a scale in India that I suppose all the practical difficulties which suggest themselves to me may be overcome.

It strikes me that it will not do to be content with a single representative of each tribe. At least four or five will be needed to eliminate the chances of accident, and even then much will depend upon the discretion and judgment of the local agent who makes the suggestion. This difficulty, however, applies chiefly if not solely to physical ethnology. To the philologer the opportunities for comparing dialects and checking pronunciation will be splendid, however [few] the individual speakers of each dialect may be. The most difficult task of all will be to prevent the assembled Savans from massacring the " specimens" at the end of the exhibition for the sake of their skulls and pelves !

I am really afraid that my own virtue might yield if so tempted !

Jesting apart, I heartily wish your plans success, and if there are any more definite ways in which I can help, let me know, and I will do my best. You will want, I should think, a physical and a philological committee to organise schemes : (1) for systematic measuring, weighing, and portraiture, with observation and recording of all physical characters ; and (2) for uniform registering of sounds by Roman letters and collection of vocabularies and grammatical forms upon an uniform system.

I should advise you to look into the Museum of the Société d'Anthropologie of Paris, and to put yourself in communication with M. Paul Broca, one of its most active members, who has lately been organising a scheme of general anthropological instructions. But don't have anything to do with the quacks who are at the head of the " Anthropological Society" over here. If they catch scent of what you are about they will certainly want to hook on to you.

Once more I wish I had the chance of being able to visit your congress. I have been lecturing on Ethnology this year,[1] and shall be again this year, and I would give a good deal to be able to look at the complex facts of Indian Ethnology with my own eyes.

But as the sage observed, " what's impossible can't be," and what with short holidays—a wife and seven children —and miles of work in arrear, India is an impossibility for me.

You say nothing about yourself, so I trust you are well and hearty, and all your belongings flourishing.— Ever yours faithfully,   T. H. HUXLEY.

---

[1] As Fullerian Professor at the Royal Institution.

In paleontology he published this year papers on the "Vertebrate Remains from the Jarrow Colliery, Kilkenny;" on a new "Telerpeton from Elgin," and on some "Dinosaurs from South Africa." The latter, and many more afterwards, were sent over by a young man named Alfred Brown, who had a curious history. A Quaker gentleman came across him when employed in cleaning tools in Cirencester College, found that he was a good Greek and Latin scholar, and got him a tutorship in a clergyman's family at the Cape. He afterwards entered the postal service, and being inspired with a vivid interest in geology, spent all the leave he could obtain from his office on the Orange River in getting fossils from the Stormberg Rocks. These, as often as he could afford to send such weighty packages, he sent to Sir R. Murchison, to whom he had received a letter of introduction from his official superior. Sir Roderick, writing to Huxley, says "that he was proud of his new recruit," to whom he sent not only welcome words of encouragement, but the no less welcome news that the brother of his "discoverer," hearing of the facts from Professor Woodward, offered to defray his expenses so that he could collect regularly.

On April 2 Huxley was in Edinburgh to receive the first academic distinction conferred upon him in Britain. He received the honorary degree of the University in company with Tyndall and Carlyle. It was part of the fitness of things that he should be associated in this honour with his close friend

Tyndall; but though he frequently acknowledged his debt to Carlyle as the teacher who in his youth had inspired him with his undying hatred of shams and humbugs of every kind, and whom he had gratefully come to know in after days, Carlyle did not forgive the publication of *Man's Place in Nature.* Years after, near the end of his life, my father saw him walking slowly and alone down the opposite side of the street, and touched by his solitary appearance, crossed over and spoke to him. The old man looked at him, and merely remarking, "You're Huxley, aren't you? the man that says we are all descended from monkeys," went on his way.

On July 6 he writes to tell Darwin that he has lodged a memorial of his about the fossils at the Gallegos river, which was to be visited by the *Nassau*[1] exploring ship, with the hydrographer direct, instead of sending it in to the Lords of the Admiralty, who would only have sent it on to the hydrographer. This letter he heads "Country orders executed with accuracy and despatch."

The following letter to Charles Kingsley explains itself—

JERMYN STREET, *April* 12, 1866.

MY DEAR KINGSLEY—I shall certainly do myself the pleasure of listening to you when you preach at the Royal Institution. I wonder if you are going to take the line of showing up the superstitions of men of science. Their name is legion, and the exploit would be a telling

---

[1] Cp. p. 449.

one. I would do it myself only I think I am already sufficiently isolated and unpopular.

However, whatever you are going to do, I am sure you will speak honestly and well, and I shall come and be assistant bottle-holder.

I am glad you like the working-men's lectures. I suspect they are about the best things of that line that I have done, and I only wish I had had the sense to anticipate the run they have had here and abroad, and I would have revised them properly.

As they stand they are terribly in the rough, from a literary point of view.

No doubt crib-biting, nurse-biting and original sin in general are all strictly deducible from Darwinian principles ; but don't by misadventure run against any academical facts.

Some whales have all the cerebral vertebræ free *now*, and every one of them has the full number, seven, whether they are free or fixed. No doubt whales had hind legs once upon a time. If when you come up to town you go to the College of Surgeons, my friend Flower the Conservator (a good man whom you should know) will show you the whalebone whale's thigh bones in the grand skeleton they have recently set up. The legs, to be sure, and the feet are gone, the battle of life having left private Cetacea in the condition of a Chelsea pensioner.—Ever yours faithfully,   T. H. HUXLEY.

This year the British Association met at Nottingham, and Huxley was president of Section D. In this capacity he invited Professor Haeckel to attend the meeting, but the impending war with Austria prevented any Prussian from leaving his country at the time, though Haeckel managed to come over later.

Huxley did not deliver a regular opening address to the section on the Thursday, but on the Friday made a speech, which was followed by a discussion upon biology and its several branches, especially morphology and its relation to physiology ("the facts concerning form are questions of force, every form is force visible.") He lamented that the sub-divisions of the section had to meet separately as a result of specialisation, the reason for which he found in the want of proper scientific education in schools. And this was the fault of the universities, for just as in the story, "Stick won't beat dog, dog won't bite pig, and so the old woman can't get home," science would not be taught in the schools until it was recognised by the universities.

This prepared the way for Dean Farrar's paper on science teaching in the public schools. His experience as a master at Harrow made him strongly oppose the existing plan of teaching all boys classical composition whether they were suited for it or no. He wished to exchange a great deal of Latin verse-making for elementary science.

This paper was doubly interesting to Huxley, as coming from a classical master in a public school, and he remarked, "He felt sure that at the present time, the important question for England was not the duration of her coal, but the due comprehension of the truths of science, and the labours of her scientific men."

On the practical side, however, Mr. J. Payne said

the great difficulty was the want of teachers; and suggested that if men of science were really in earnest they would condescend to teach in the schools.

It was to a certain extent in answer to this appeal that Huxley gave his lectures on Physiography in 1869 (see p. 444), and instituted the course of training for science teachers in 1871.

He concluded his work at Nottingham by a lecture to working men.

The following is in reply to Mr. Spencer, who had accused himself of losing his temper in an argument—

26 ABBEY PLACE, *Sunday, Nov.* 8, 1866.

MY DEAR SPENCER—Your conscience has been treating you with the most extreme and unjust severity.

I recollect you *looked* rather savage at one point in our discussion, but I do assure you that you committed no overt act of ferocity; and if you had, I think I should have fully deserved it for joining in the ferocious onslaught we all made upon you.

What your sins may be in this line to other folk I don't know, but so far as I am concerned I assure you I have often said that I know no one who takes aggravated opposition better than yourself, and that I have not a few times been ashamed of the extent to which I have tried your patience.

So you see that you have, what the Buddhists call a stock of accumulated merit, *envers moi*—and if you should ever feel inclined to "d—n my eyes" you can do so and have a balance left.

Seriously, my old friend, you must not think it necessary to apologise to me about any such matters, but believe me (d—ned or und—d)—Ever yours faithfully,

T. H. HUXLEY.

26 ABBEY PLACE, *Nov.* 11, 1866.

MY DEAR DARWIN—I thank you for the new edition of the *Origin*, and congratulate you on having done with it for a while, so as to be able to go on to that book of a portion of which I had a glimpse years ago. I hear good accounts of your health, indeed the last was that you were so rampageous you meant to come to London and have a spree among its dissipations. May that be true.

I am in the thick of my work, and have only had time to glance at your *Historical Sketch*.

What an unmerciful basting you give "our mutual friend." I did not know he had put forward any claim! and even now that I read it in black and white, I can hardly believe it.

I am glad to hear from Spencer that you are on the right (that is *my*) side in the Jamaica business. But it is wonderful how people who commonly act together are divided about it.

My wife joins with me in kindest wishes to Mrs. Darwin and yourself—Ever yours faithfully,

T. H. HUXLEY.

You will receive an elementary physiology book, not for your reading but for Miss Darwin's. Were you not charmed with Haeckel?

The "Jamaica business" here alluded to was Governor Eyre's suppression of a negro rising, in the course of which he had executed, under martial law, a coloured leader and member of the Assembly, named Gordon. The question of his justification in so doing stirred England profoundly. It became the touchstone of ultimate political convictions. Men who had little concern for ordinary politics came forward to defend a great constitutional principle which they conceived to be endangered. A committee was formed to prosecute Governor Eyre on a charge of murder, in order to vindicate the right of a prisoner to trial by due process of law. Thereupon a counter-committee was organised for the defence of the man who, like Cromwell, judged that the people preferred their real security to forms, and had presumably saved the white population of Jamaica by striking promptly at the focus of rebellion.

The *Pall Mall Gazette* of October 29, 1866, made a would-be smart allusion to the part taken in the affair by Huxley, which evoked, in reply, a calm statement of his reasons for joining the prosecuting committee :—

It is amusing (says the *Pall Mall*) to see how the rival committees, the one for the prosecution and the other for the defence of Mr. Eyre, parade the names of distinguished persons who are enrolled as subscribers on either side. Mill is set against Carlyle, and to counterbalance the adhesion of the Laureate to the Defence Fund, the *Star* hastens to announce that Sir Charles Lyell and Professor Huxley have given their support to the Jamaica Committee. Everything, of course, depends on the ground on which the subscriptions are given. One can readily conceive that Mr. Tennyson has been chiefly moved by a generous indignation at the vindictive behaviour of the Jamaica Committee. It would be curious also to know how far Sir Charles Lyell's and Mr. Huxley's peculiar views on the development of species have influenced them in bestowing on the negro that sympathetic recognition which they are willing to extend even to the ape as "a man and a brother."

The reply appeared in the *Pall Mall* of October 31 :—

SIR—I learn from yesterday evening's *Pall Mall Gazette* that you are curious to know whether certain "peculiar views on the development of species," which I am said to hold in the excellent company of Sir Charles Lyell, have led me to become a member of the Jamaica Committee.

Permit me without delay to satisfy a curiosity which does me honour. I have been induced to join that committee neither by my "peculiar views on the development of species," nor by any particular love for, or admiration of the negro—still less by any miserable desire to wreak vengeance for recent error upon a man whose early career I have often admired ; but because the course which the committee proposes to take appears to me to be the only one by which a question of the profoundest practical importance can be answered. That question is, Does the killing a man in the way Mr. Gordon was killed constitute murder in the eye of the law, or does it not ?

You perceive that this question is wholly independent of two others which are persistently confused with it, namely—was Mr. Gordon a Jamaica Hampden or was he a psalm-singing fire-brand ? and was Mr. Eyre actuated by the highest and noblest motives, or was he under the influence of panic-stricken rashness or worse impulses ?

I do not presume to speak with authority on a legal question ; but, unless I am misinformed, English law does not permit good persons, as such, to strangle bad persons, as such. On the contrary, I understand that, if the most virtuous of Britons, let his place and authority be what they may, seize and hang up the greatest scoundrel in her Majesty's dominions simply because he is an evil and troublesome person, an English court of justice will certainly find that virtuous person guilty of murder. Nor will the verdict be affected by any evidence that the defendant acted from the best of motives, and, on the whole, did the State a service.

Now, it *may* be that Mr. Eyre was actuated by the best of motives ; it *may* be that Jamaica is all the better for being rid of Mr. Gordon ; but nevertheless the Royal Commissioners, who were appointed to inquire into Mr. Gordon's case, among other matters, have declared that :—

The evidence, oral and documentary, appears to us to be wholly insufficient to establish the charge upon which the prisoner took his trial. (*Report*, p. 37.)

And again that they

Cannot see in the evidence which has been adduced any sufficient proof, either of his (Mr. Gordon's) complicity in the outbreak at Morant Bay, or of his having been a party to any general conspiracy against the Government. (*Report*, p. 38.)

Unless the Royal Commissioners have greatly erred, therefore, the killing of Mr. Gordon can only be defended on the ground that he was a bad and troublesome man ; in short, that although he might not be guilty, it served him right.

I entertain so deeply-rooted an objection to this method of killing people—the act itself appears to me to be so frightful a precedent, that I desire to see it stigmatised by the highest authority as a crime. And I have joined the committee which proposes to indict Mr. Eyre, in the hope that I may hear a court of justice declare that the only defence which can be set up (if the Royal Commissioners are right) is no defence, and that the killing of Mr. Gordon was the greatest offence known to the law— murder.—I remain, Sir, your obedient servant,

THOMAS H. HUXLEY.

THE ATHENÆUM CLUB, *Oct.* 30, 1866.

Two letters to friends who had taken the opposite

side in this burning question show how resolutely he set himself against permitting a difference on matters of principle to affect personal relations with his warmest opponents.

JERMYN STREET, *Nov.* 8, 1866.

MY DEAR KINGSLEY—The letter of which you have heard, containing my reasons for becoming a member of the Jamaica Committee, was addressed to the *Pall Mall Gazette* in reply to some editorial speculations as to my reasons for so doing.

I forget the date of the number in which my letter appeared, but I will find it out and send you a copy of the paper.

Mr. Eyre's personality in this matter is nothing to me; I know nothing about him, and, if he is a friend of yours, I am very sorry to be obliged to join in a movement which must be excessively unpleasant to him.

Furthermore, when the verdict of the jury which will try him is once given, all hostility towards him on my part will cease. So far from wishing to see him vindictively punished, I would much rather, if it were practicable, indict his official hat and his coat than himself.

I desire to see Mr. Eyre indicted and a verdict of guilty in a criminal court obtained, because I have, from its commencement, carefully watched the Gordon case; and because a new study of all the evidence which has now been collected has confirmed. my first conviction that Gordon's execution was as bad a specimen as we have had since Jeffreys' time of political murder.

Don't suppose that I have any particular admiration for Gordon. He belongs to a sufficiently poor type of small political agitator—and very likely was a great nuisance to the Governor and other respectable persons. But that is no reason why he should be condemned, by an absurd tribunal and with a brutal mockery of the forms of justice, for offences with which impartial judges, after a full investigation, declare there is no evidence to show that he was connected.

Ex-Governor Eyre seized the man, put him in the hands of the preposterous subalterns, who pretended to try him—saw the evidence and approved of the sentence. He is as much responsible for Gordon's death as if he had shot him through the head with his own hand. I daresay he did all this with the best of motives, and in a heroic vein. But if English law will not declare that heroes have no more right to kill people in this fashion than other folk, I shall take an early opportunity of migrating to Texas or some other quiet place where there is less hero-worship and more respect for justice, which is to my mind of much more importance than hero-worship.

In point of fact, men take sides on this question, not so much by looking at the mere facts of the case, but rather as their deepest political convictions lead them. And the great use of the prosecution, and one of my reasons for joining it, is that it will help a great many people to find out what their profoundest political beliefs are.

The hero-worshippers who believe that the world is to be governed by its great men, who are to lead the little ones, justly if they can ; but if not, unjustly drive or kick them the right way, will sympathise with Mr. Eyre.

The other sect (to which I belong) who look upon hero-worship as no better than any other idolatry, and upon the attitude of mind of the hero-worshipper as essentially immoral ; who think it is better for a man to go wrong in freedom than to go right in chains ; who look upon the observance of inflexible justice as between man and man as of far greater importance than even the preservation of social order, will believe that Mr. Eyre has committed one of the greatest crimes of which a person in authority can be guilty, and will strain every nerve to

obtain a declaration that their belief is in accordance with
the law of England.

People who differ on fundamentals are not likely to
convert one another. To you, as to my dear friend
Tyndall, with whom I almost always act, but who in this
matter is as much opposed to me as you are, I can only
say, let us be strong enough and wise enough to fight the
question out as a matter of principle and without bitterness.—Ever yours faithfully,   T. H. HUXLEY.

November 9, 1866.

MY DEAR TYNDALL—Many thanks for the kind note
which accompanied your letter to the Jamaica Committee.

When I presented myself at Rogers' dinner last night
I had not heard of the latter, and Gassiot began poking
fun at me, and declaring that your absence was due to a
quarrel between us on this unhappy subject.

I replied to the jest earnestly enough, that I hoped
and believed our old friendship was strong enough to
stand any strain that might be put on it, much as I
grieved that we should be ranged in opposite camps in
this or any other cause.

That you and I have fundamentally different political
principles must, I think, have become obvious to both of
us during the progress of the American War. The fact
is made still more plain by your printed letter, the tone
and spirit of which I greatly admire without being able
to recognise in it any important fact or argument which
had not passed through my mind before I joined the
Jamaica Committee.

Thus there is nothing for it but for us to agree to
differ, each supporting his own side to the best of his
ability, and respecting his friend's freedom as he would
his own, and doing his best to remove all petty bitterness
from that which is at bottom one of the most important
constitutional battles in which Englishmen have for many
years been engaged.

If you and I are strong enough and wise enough, we shall be able to do this, and yet preserve that love for one another which I value as one of the good things of my life.

If not, we shall come to grief.　I mean to do my best. —Ever yours faithfully,　　　　　　T. H. HUXLEY.

Governor Eyre, however, though recalled, escaped condemnation on the ground that he was honestly convinced of Gordon's guilt and had acted without legal "malice." The Jamaica Committee raised the question in Parliament, but the Government refused to prosecute him for murder, though ready to revise the sentences of those he had punished and to give them compensation. In these circumstances the House contented itself with a resolution deploring the excessive punishments, and the subsequent efforts of the Jamaica Committee to obtain a conviction in a court of law were fruitless. The magistrates of Shropshire, where Eyre was then living, refused to issue a warrant against the ex-Governor; and in London, grand juries ignored the bills against Colonel Nelson and one of the "preposterous subalterns" in 1867, and against Eyre himself in 1868. The latter spent the remainder of his life in retirement, dying in November 1901 at the age of 86.

Huxley was always of opinion that to write a good elementary text-book required a most extensive and intimate knowledge of the subject under discussion. Certainly the *Lessons on Elementary Physiology* which appeared at the end of 1866 were the outcome of such knowledge, and met with a wonderful and lasting success as a text-book. A graceful compliment was passed upon it by Sir William Lawrence, when, in thanking the author for the gift of the book, he wrote (January 24, 1867), "in your modest book 'indocti discant, ament meminisse periti!'"

This was before the days of American copyright, and English books were usually regarded as fair prey by the mass of American publishers. Among the exceptions to this practical rule were the firm of Appletons, who made it a point of honour to treat foreign authors as though they were legally entitled to some equitable rights. On their behalf an arrangement was made for an authorised American edition of the *Physiology* by Dr. Youmans, whose acquaintance thus made my father did not allow to drop.

It is worth noting that by the year 1898 this little book had passed through four editions, and been reprinted thirty-one times.

CHAPTER XXI

1867

It has already been noted that Huxley's ethnological work continued this year with a second series of lectures at the Royal Institution, while he enlarged his paper on "Two widely contrasted forms of Human Crania," and published it in the *Journal of Anatomy*. One paleontological memoir of his appeared this year on Acanthopholis, a fossil from the chalk marl, an additional piece of work for which he excuses himself to Sir C. Lyell (January 4, 1867):—

The new reptile advertised in *Geol. Mag.* has turned up in the way of business, and I could not help giving a notice of it, or I should not have undertaken anything fresh just now.

The Spitzbergen things are very different, and I have taken sundry looks at them and put them by again to let my thoughts ripen.

They are Ichthyosaurian, and I am not sure they do not belong to two species. But it is an awful business to compare all the Ichthyosaurians. I *think* that one form is new. Please to tell Nordenskiold this much.

However, his chief interest was in the anatomy of birds, at which he had been working for some time, and especially the development of certain of the cranial bones as a basis of classification. On April 11, expanding one of his Hunterian Lectures, he read a paper on this subject at the Zoological Society, afterwards published in their *Proceedings* for 1867.

As he had found the works of Professor Cornay of help in the preparation of this paper, he was careful to send him a copy with an acknowledgment of his indebtedness, eliciting the reply "*c'est si beau de trouver chez l'homme la science unie à la justice.*"

He followed this up with another paper on "The Classification and Distribution of the Alectoromorphæ and Heteromorphæ" in 1868, and to the work upon this the following letter to his ally, W. K. Parker, refers :—

ROYAL GEOLOG. SURVEY OF GT. BRITAIN,
JERMYN STREET, *July* 17, 1867.

MY DEAR PARKER—Nothing short of the direct temptation of the evil one could lead you to entertain so monstrous a doctrine, as that you propound about *Cariamidæ.*

I recommend fasting for three days and the application of a scourge thrice in the twenty-four hours ! Do this, and about the fourth day you will perceive that the cranial differences alone are as great as those between *Cathartes* and *Serpentarius.*

If you want to hear something new and true it is this :—

1. That Memora is more unlike all the other Passerines

(*i.e.* Coracomorphæ) than they are unlike one another, and that it will have to stand in a group by itself.

It is as much like a wren as you are—less so, in fact, if you go on maintaining that preposterous fiction about Serpentarius.

2. Wood-peckers are more like crows than they are like cuckoos.

Aegithognathæ

Coracomorphæ

Cypselomorphæ   Gecinomorphæ

Desmognathæ

Coccygomorphæ.

3. Sundevell is the sharpest fellow who has written on the classification of birds.

4. Nitzsch and W. K. Parker[1] are the sharpest fellows who have written on their osteology.

5. Though I do not see how it follows naturally on the above, still, where can I see a good skeleton of Glareola?

None in college, B.M.S. badly prepared.—Ever yours faithfully,   T. H. HUXLEY.

[1] Except in the case of Serpentarius.

An incident which diversified one of the Gilchrist lectures to working men is thus recorded by the *Times* of January 23, 1867 :—

A GOOD EXAMPLE. Last night, at the termination of a lecture on ethnology, delivered by Professor Huxley to an audience which filled the theatre of the London Mechanics' Institute in Southampton Buildings, Chancery Lane, the lecturer said that he had received a letter as he entered the building which he would not take the responsibility of declining to read, although it had no reference to the subject under consideration. He then read the letter, which was simply signed "A Regular Attendant at Your Lectures," and which in a few words drew attention to the appalling distress existing among the population out of work at the East End, and suggested that all those present at the lecture that night should be allowed the opportunity of contributing 1d. or 2d. each towards a fund for their relief, and that the professor should become the treasurer for the evening. This suggestion was received by the audience with marks of approval. The professor said he would not put pressure on anyone ; he would simply place his own subscription in one of the skulls on the table. This he did, and all the audience coming on the platform, threw in money in copper and silver until the novel cash-box was filled with coin which amounted to a large sum. A gentleman present expressed a hope that the example set by that audience might be followed with good results wherever large bodies assembled either for educational or recreative purposes.

At the end of April this year my father spent a week in Brittany with Dr. Hooker and Sir J. Lubbock, rambling about the neighbourhood of Rennes and Vannes, and combining the examination of prehistoric remains with the refreshment of holiday-making.

Few letters of this period exist. The x Club was doing its work. Most of those to whom he would naturally have written he met constantly. Two letters to Professor Haeckel give pieces of his experience. One suggests the limits of aggressive polemics, as to which I remember his once saying that he himself had only twice been the aggressor in controversy, without waiting to be personally attacked; once where he found his opponent was engaged in a flanking movement; the other when a man of great public reputation had come forward to champion an untenable position of the older orthodoxy, and a blow dealt to his pretensions to historical and scientific accuracy would not only bring the question home to many who neglected it in an impersonal form, but would also react upon the value of the historical arguments with which he sought to stir public opinion in other spheres. The other letter touches on the influence, at once calming and invigorating, as he had known it to the full for the last twelve years, which a wife can bring in the midst of outward struggles to the inner life of the home.

JERMYN STREET, LONDON, *May* 20, 1867.

MY DEAR HAECKEL—Your letter, though dated the 12th, has but just reached me. I mention this lest you should think me remiss, my sin in not writing to you already being sufficiently great. But your book did not reach me until November, and I have been hard at work lecturing, with scarcely an intermission ever since

Now I need hardly say that the *Morphologie* is not exactly a novel to be taken up and read in the intervals of business. On the contrary, though profoundly interesting, it is an uncommonly hard book, and one wants to read every sentence of it over.

I went through it within a fortnight of its coming into my hands, so as to get at your general drift and purpose, but up to this time I have not been able to read it as I feel I ought to read it before venturing upon criticism. You cannot imagine how my time is frittered away in these accursed lectures and examinations.

There can be but one opinion, however, as to the knowledge and intellectual grasp displayed in the book; and, to me, the attempt to systematise biology as a whole is especially interesting and valuable.

I shall go over this part of your work with great care by and by, but I am afraid you must expect that the number of biologists who will do so, will remain exceedingly small. Our comrades are not strong in logic and philosophy.

With respect to the polemic *excursus*, of course, I chuckle over them most sympathetically, and then say how naughty they are! I have done too much of the same sort of thing not to sympathise entirely with you; and I am much inclined to think that it is a good thing for a man, once at any rate in his life, to perform a public war-dance against all sorts of humbug and imposture.

But having satisfied one's love of freedom in this way, perhaps the sooner the war-paint is off the better. It has no virtue except as a sign of one's own frame of mind and determination, and when that is once known, is little better than a distraction.

I think there are a few patches of this kind, my dear friend, which may as well come out in the next edition, *e.g.* that wonderful note about the relation of God to gas, the gravity of which greatly tickled my fancy.

I pictured to myself the effect which a translation of this would have upon the minds of my respectable countrymen!

Apropos of translation. Darwin wrote to me on that subject, and with his usual generosity, would have made à considerable contribution towards the expense if we could have seen our way to the publication of a translation. But I do not think it would be well to translate the book in fragments, and, as a whole, it would be a very costly undertaking, with very little chance of finding readers.

I do not believe that in the British Islands there are fifty people who are competent to read the book, and of the fifty, five and twenty have read it or will read it in German.

What I desire to do is to write a review of it, which will bring it into some notice on this side of the water, and this I hope to do before long. If I do not, it will be, you well know, from no want of inclination, but simply from lack of time.

In any case, as soon as I have been able to study the book carefully, you shall have my honest opinion about all points.

I am glad your journey has yielded so good a scientific harvest, and especially that you found my *Oceanic Hydrozoa* of some use. But I am shocked to find you had no copy of the book of your own, and I shall take care that one is sent to you. It is my firstborn work, done when I was very raw and inexperienced, and had neither friends nor help. Perhaps I am all the fonder of the child on that ground.

A lively memory of you remains in my house, and wife and children will be very glad to hear that I have news of you when I go home to dinner.

Keep us in kindly recollection, and believe me—Ever yours very faithfully,   T. H. HUXLEY.

July 16, 1867.

MY DEAR HAECKEL—My wife and I send you our most hearty congratulations and good wishes. Give your betrothed a good account of us, for we hope in the future to entertain as warm a friendship for her as for you. I was very glad to have the news, for it seemed to me very sad that a man of your warm affections should be surrounded only by hopeless regrets. Such surroundings inflict a sort of partial paralysis upon one's whole nature, a result which is, to me, far more serious and regrettable than the mere suffering one undergoes.

The one thing for men, who like you and I stand pretty much alone, and have a good deal of fighting to do in the external world, is to have light and warmth and confidence within the four walls of home. May all these good things await you!

Many thanks for your kind invitation to Jena. I am sure my wife would be as much pleased as I to accept it, but it is very difficult for her to leave her children.

We will keep it before us as a pleasant possibility, but I suspect you and Madame will be able to come to England before we shall reach Germany.

I wish I had rooms to offer you, but you have seen that troop of children, and they leave no corner unoccupied.

Many thanks for the Bericht and the genealogical tables. You seem, as usual, to have got through an immense amount of work.

I have been exceedingly occupied with a paper on the "Classification of Birds," a sort of expansion of one of my Hunterian Lectures this year. It has now gone to press, and I hope soon to be able to send you a copy of it.

Occupation of this and other kinds must be my excuse for having allowed so much longer a time to slip by than I imagined had done before writing to you. It is not for want of sympathy, be sure, for my wife and I have often talked of the new life opening out to you.

This is written in my best hand. I am proud of it, as I can read every word quite easily myself, which is more than I can always say for my own MS.—Ever yours faithfully,   T. H. HUXLEY.

The same experience is attested and enforced in the correspondence with Dr. Anton Dohrn, which begins this year. Genial, enthusiastic, as pungent as he was eager in conversation, the future founder of the Marine Biological Station at Naples, on his first visit to England, made my father's acquaintance by accepting his invitation to stay with him "for as long as you can make it convenient to stay" at Swanage, "a little country town with no sort of amusement except what is to be got by walking about a rather pretty country. But having warned you of this, I repeat that it will give me much pleasure to see you if you think it worth while to come so far."

Dr. Dohrn came, and came into the midst of the family—seven children, ranging from ten years to babyhood, with whom he made himself as popular by his farmyard repertory, as he did with the elders by other qualities. The impression left upon him appears from a letter written soon after—

"Ich habe heute mehrere Capitel in Mill's *Utilitarianism* gelesen und das Wort happiness mehr als einmal gefunden : hatte *ich* eine Definition dieses vielumworbenen Wortes irgend Jemand zu geben, ich würde sagen :[1] go and see the Huxley family at Swanage ; and if you would enjoy the same I enjoyed, you would feel what is happiness, and never more ask for a definition of this sentiment."

SWANAGE, *Sept.* 22, 1867.

MY DEAR DOHRN—Thanks to my acquaintance with the *Mikroskopische Anatomie,* and to the fact that you employ our manuscript characters, and not the hieroglyphics of what I venture to call the "cursed" and not "cursiv" Schrift, your letter was as easy as it was pleasant to read. We are all glad to have news of you, though

---

[1] I have been reading several chapters of Mill's *Utilitarianism* to-day, and met with the word "happiness" more than once ; if *I* had to give anybody a definition of this much debated word, I should say—

it was really very unnecessary to thank us for trying to make your brief visit a pleasant one. Your conscience must be more "pungent" than your talk, if it pricks you with so little cause. My wife rejoices saucily to find that phrase of hers has stuck so strongly in your mind, but you must remember her fondness for "Tusch."

You must certainly marry. In my bachelor days, it was unsafe for anyone to approach me before mid-day, and for all intellectual purposes I was barren till the evening. Breakfast at six would have upset me for the day. You and the lobster noted the difference the other day.

Whether it is matrimony or whether it is middle age I don't know, but as time goes on you can combine both.

I cannot but accept your kind offer to send me Fanny Lewald's works, though it is a shame to rob you of them. In return my wife insists on your studying a copy of Tennyson, which we shall send you as soon as we return to civilisation, which will be next Friday. If you are in London after that date, we shall hope to see you once more before you return to the bosom of the "Father-land."

I did my best to give the children your message, but I fear I failed ignominiously in giving the proper bovine vocalisation to "Mroo."

That small curly-headed boy Harry, struck, I suppose by the kindness you both show to children, has effected a synthesis between you and Tyndall, and gravely observed the other day, "Doctor Dohrn-Tyndall do say Mroo."

My wife sends her kind regards. The "seven" are not here or they would vote love by acclamation.—
Ever yours very faithfully,   T. H. HUXLEY.

He did not this year attend the British Associa-
tion, which was held in Dundee. This was the first occasion on which an evening was devoted to a

working-men's lecture, a step important as tending towards his own ideal of what science should be :— not the province of the few, but the possession of the many.

This first lecture was delivered by Professor Tyndall, who wrote him an account of the meeting, and in particular of his reconciliation with Professors Thomson (Lord Kelvin) and Tait, with whom he had had a somewhat embittered controversy.

In his reply, Huxley writes :—

To J. Tyndall

Thanks also for a copy of the *Dundee Advertiser* containing your lecture. It seemed to me that the report must be a very good one, and the lecture reads exceedingly well. You have inaugurated the working-men's lectures of the Association in a way that cannot be improved. And it was worth the trouble, for I suspect they will become a great and noble feature in the meetings.

Everything seems to have gone well at the meeting, the educational business carried [*i.e.* a recommendation that natural science be made a part of the curriculum in the public schools], and the anthropologers making fools of themselves in a most effectual way. So that I do not feel I have anything to reproach myself with for being absent.

I am very pleased to hear of the reconciliation with Thomson and Tait. The mode of it speaks well for them, and the fact will remove a certain source of friction from amongst the cogs of your mental machinery.

The following gives the reason for his resigning the Fullerian lectureship :—

ATHENÆUM CLUB, *May*, 1867.

MY DEAR TYNDALL—A conversation I had with
Bence Jones yesterday reminded me that I ought to have
communicated with you. But we do not meet so often
as we used to do, being, I suppose, both very busy, and
I forget to write.

You recollect that the last time we talked together,
you mentioned a notion of Bence Jones's to make the
Fullerian Professorship of Physiology a practically permanent appointment, and that I was quite inclined to
stick by that (if such arrangement could be carried out),
and give up other things.

But since I have been engaged in the present course
of lectures I have found reason to change my views. It
is very hard work, and takes up every atom of my time
to make the lectures what they should be ; and I find
that at this time of year, being more or less used up, I
suppose, with the winter work, I stand the worry and
excitement of the actual lectures very badly. Add to
this that it is six weeks clean gone out of the only time
I have disposable for real scientific progress, and you
will understand how it is that I have made up my mind
to resign.

I put all this clearly before Bence Jones yesterday,
with the proviso that I could and would do nothing that
should embarrass the Institution or himself.

If there is the least difficulty in supplying my place,
or if the managers think I shall deal shadily with them
by resigning before the expiration of my term, of course
I go on. And I hope you all understand that I would
do anything rather than put even the appearance of a
slight upon those who were kind enough to elect me.—
Ever yours,   T. H. HUXLEY.

He found a substitute for 1868, the last year of
the triennial course, in Dr. (now Sir) Michael Foster.

Of his final lectures in 1867 he used to tell a story against himself.

In my early period as a lecturer, I had very little confidence in my general powers, but one thing I prided myself upon was clearness. I was once talking of the brain before a large mixed audience, and soon began to feel that no one in the room understood me. Finally I saw the thoroughly interested face of a woman auditor, and took consolation in delivering the remainder of the lecture directly to her. At the close, my feeling as to her interest was confirmed when she came up and asked if she might put one question upon a single point which she had not quite understood. "Certainly," I replied. "Now, Professor," she said, "is the cerebellum inside or outside the skull?" (*Reminiscences of T. H. Huxley*, by Professor H. Fairfield Osborn).

Dr. Foster used to add maliciously, that disgust at the small impression he seemed to have made was the true reason for the transference of the lectures.

CHAPTER XXII

1868

IN 1868 he published five scientific memoirs, amongst them his classification of birds and "Remarks upon Archæopteryx Lithographica" (*Proc. Roy. Soc.* xvi. 1868, pp. 243-248). This creature, a bird with reptilian characters, was a suggestive object from which to popularise some of the far-reaching results of his many years' labour upon the morphology of both birds and reptiles. Thus it led to a lecture at the Royal Institution, on February 7, "On the Animals which are most nearly intermediate between Birds and Reptiles.'

Of this branch of work Sir M. Foster says (Obit. Not. *Proc. Roy. Soc.* vol. lix.) :—

One great consequence of these researches was that science was enriched by a clear demonstration of the many and close affinities between reptiles and birds, so that the two henceforward came to be known under the joint title of Sauropsida, the amphibia being at the same time distinctly more separated from the reptiles, and their relations to fishes more clearly signified by the joint title of Ichthyopsida. At the same time, proof was brought forward that the line of descent of the Sauropsida clearly diverged from that of the Mammalia, both starting from some common ancestry. And besides this great generalisation, the importance of which, both from a classificatory and from an evolutional point of view, needs no comment, there came out of the same researches numerous lesser contributions to the advancement of morphological knowledge, including among others an attempt, in many respects successful, at a classification of birds.

This work in connection with the reptilian ancestry of birds further appears in the paleontological papers published in 1869 upon the Dinosaurs (see Chap. XXIII.), and is referred to in a letter to Haeckel, p. 437.

His Hunterian lectures on the Invertebrata appeared this year in the *Quarterly Journal of Microscopical Science* (pp. 126-129, and 191-201), and in the October number of the same journal appeared his famous article "On some Organisms living at great depth in the North Atlantic Ocean," originally delivered before the British Association at Norwich in this year (1868). The sticky or viscid character of the fresh mud from the bottom of the Atlantic had already been noticed by Captain Dayman when making soundings for the Atlantic cable. This stickiness was apparently due to the presence of innumerable lumps of a transparent, gelatinous substance consisting of minute granules without discoverable nucleus or membranous envelope, and interspersed with calcareous coccoliths. After a description of the structure of this substance and its chemical reactions, he makes a careful proviso against confounding the statement of fact in the description and the interpretation which he proceeds to put upon these facts :—

I conceive that the granulate heaps and the transparent gelatinous matter in which they are embedded represent masses of protoplasm. Take away the cysts which characterise the *Radiolaria*, and a dead *Sphaerozoum* would very nearly represent one of this deep-sea "Urschleim," which must, I think, be regarded as a new form of those simple animated beings which have recently been so well described by Haeckel in his *Monographie der Moneras*, p. 210.[1]

Of this he writes to Haeckel on October 6, 1868 :—

[This paper] is about a new "Moner" which lies at the bottom of the Atlantic to all appearances, and gives rise to some wonderful calcified bodies. I have christened it *Bathybius Haeckelii*, and I hope that you will not be ashamed of your god-child. I will send you some of the mud with the paper.

The explanation was plausible enough on general grounds, if the evidence had been all that it seemed to be. But it must be noted that the specimens examined by him and by Haeckel, who two years later published a full and detailed description of Bathybius, were seen in a preserved state. Neither of them saw a fresh specimen, though on the cruise of the *Porcupine*, Sir Wyville Thomson and Dr. W. B. Carpenter examined the substance in a fresh state,

[1] See *Coll. Ess.* v. 153.

and found no better explanation to give of it. However, not only were the expectations that it was very widely distributed over the Atlantic bottom falsified in 1879 by the researches of the *Challenger* expedition, but the behaviour of certain deep-sea specimens gave good ground for suspecting that what had been sent home before as genuine deep-sea mud was a precipitate due to the action on the specimens of the spirit in which they were preserved. Though Haeckel, with his special experience of Monera, refused to desert Bathybius, a close parallel to which was found off Greenland in 1876, the rest of its sponsors gave it up. Whatever it might be as a matter of possibility, the particular evidence upon which it had been described was tainted. Once assured of this, Huxley characteristically took the bull by the horns. Without waiting for any one else to come forward, he made public renunciation of Bathybius at the British Association in 1879.[1] The "eating of the leek" as recommended to his friend Dohrn (July 7, 1868), was not merely a counsel for others, but was a prescription followed by himself on occasion : —

As you know, I did not think you were on the right track with the Arthropoda, and I am not going to profess to be sorry that you have finally worked yourself to that conclusion.

As to the unlucky publication in the *Journal of Anatomy and Physiology*, you have read your Shakespeare

---

[1] See vol. ii. p. 268, *sq.*

and know what is meant by "eating a leek." Well,
every honest man has to do that now and then, and I
assure you that if eaten fairly and without grimaces, the
devouring of that herb has a very wholesome cooling
effect on the blood, particularly in people of sanguine
temperament.

Seriously you must not mind a check of this kind.

This incident, one may suspect, was in his mind
when he wrote in his *Autobiography* of the rapidity
of thought characteristic of his mother :—

That characteristic has been passed on to me in full
strength ; it has often stood me in good stead, it has
sometimes played me sad tricks, and it has always been a
danger.

At the Norwich meeting of the Association he also
delivered his well-known lecture to working men
"On a Piece of Chalk," a perfect example of the
handling of a common and trivial subject, so as to
make it "a window into the Infinite." He was
particularly interested in the success of the meeting,
as his friend Hooker was President, and writes to
Darwin, September 12 :—

We had a capital meeting at Norwich, and dear old
Hooker came out in great force as he always does in
emergencies.

The only fault was the terrible "Darwinismus" which
spread over the section and crept out when you least
expected it, even in Fergusson's lecture on "Buddhist
Temples."

You will have the rare happiness to see your ideas
triumphant during your lifetime.

P.S.—I am preparing to go into opposition ; I can't
stand it.

This lecture "On a Piece of Chalk," together with two others delivered this year, seem to me to mark the maturing of his style into that mastery of clear expression for which he deliberately laboured, the saying exactly what he meant, neither too much nor too little, without confusion and without obscurity. Have something to say, and say it, was the Duke of Wellington's theory of style; Huxley's was to say that which has to be said in such language that you can stand cross-examination on each word. Be clear, though you may be convicted of error. If you are clearly wrong, you will run up against a fact some time and get set right. If you shuffle with your subject, and study chiefly to use language which will give a loophole of escape either way, there is no hope for you.

This was the secret of his lucidity. In no one could Buffon's aphorism on style find a better illustration, *Le style c'est l'homme même.* In him science and literature, too often divorced, were closely united; and literature owes him a debt for importing into it so much of the highest scientific habit of mind; for showing that truthfulness need not be bald, and that real power lies more in exact accuracy than in luxuriance of diction. Years after, no less an authority than Spedding, in a letter upon the influence of Bacon on his own style in the matter of exactitude, the pruning of fine epithets and sweeping statements, the reduction of numberless superlatives to positives, asserted that, if as a young man he had fallen in with Huxley's writings before Bacon's, they would have produced the same effect upon him.[1]

Of the other two discourses referred to, one is the opening address which he delivered as Principal at the South London Working Men's College on January 4, "A Liberal Education, and Where to Find It." This is not a brief for science to the exclusion of other teaching; no essay has insisted more strenuously on the evils of a one-sided education, whether it be classical or scientific; but it urged the necessity for a strong tincture of science and her method, if the modern conception of the world, created by the spread of natural knowledge, is to be fairly understood. If culture is based on knowledge of the best that has been written and thought in the world, it is incomplete without knowledge of the most important factor which has transformed the medieval into the modern spirit.

Two of his most striking passages are to be found in this address; one the simile of the force behind nature as the hidden chess-player; the other the noble description of the end of a true education.

Well known as it is, I venture to quote the latter as an instance of his style :—

That man, I think, has had a liberal education, who has been so trained in youth that his body is the ready servant of his will, and does with ease and pleasure all the work that as a mechanism it is capable of; whose intellect is a clear cold logic engine, with all its parts of

---

[1] See vol. ii. p. 239.

equal strength, and in smooth working order ; ready, like a steam engine, to be turned to any kind of work, and spin the gossamers as well as forge the anchors of the mind ; whose mind is stored with a knowledge of the great and fundamental truths of nature and of the laws of her operations ; one who, no stunted ascetic, is full of life and fire, but whose passions are trained to come to heel by a vigorous will, the servant of a tender conscience ; who has learned to love all beauty, whether of nature or of art, to hate all vileness, and to respect others as himself.

Such an one and no other, I conceive, has had a liberal education, for he is, as completely as a man can be, in harmony with nature.   He will make the best of her, and she of him.   They will get on together rarely ; she as his ever-beneficent mother ; he as her mouth-piece, her conscious self, her minister and interpreter.

The third of these discourses is the address "On the Physical Basis of Life," of which he writes to Haeckel on January 20, 1869 :—

You will be amused to hear that I went to the holy city, Edinburgh itself, the other day, for the purpose of giving the first of a series of Sunday lectures.   I came back without being stoned ; but Murchison (who is a Scotchman, you know) told me he thought it was the boldest act of my life.   The lecture will be published in February, and I shall send it to you, as it contains a criticism of materialism which I should like you to consider.

In it he explains in popular form a striking generalisation of scientific research, namely, that whether in animals or plants, the structural unit of the living body is made up of similar material, and that vital action and even thought are ultimately

based upon molecular changes in this life-stuff.
Materialism ! gross and brutal materialism ! was the
mildest comment he expected in some quarters; and
he took the opportunity to explain how he held
" this union of materialistic terminology with the
repudiation of materialistic philosophy," considering
the latter " to involve grave philosophic error."

His expectations were fully justified ; in fact, he
writes that some persons seemed to imagine that he
had invented protoplasm for the purposes of the
lecture.

Here, too, in the course of a reply to Archbishop
Thomson's confusion of the spirit of modern thought
with the system of M. Comte, he launched his well-
known definition of Comtism as Catholicism *minus*
Christianity, which involved him in a short controversy with Mr. Congreve (see "The Scientific
Aspects of Positivism," *Lay Sermons*, p. 162), and
with another leading Positivist, who sent him a
letter through Mr. Darwin.   Huxley replied :—

JERMYN STREET, *March* 11, 1869.

MY DEAR DARWIN—I know quite enough of Mr.
—— to have paid every attention to what he has to say,
even if you had not been his ambassador.

I glanced over his letter when I returned home last
night very tired with my two nights' chairmanship at the
Ethnological and the Geological Societies.

Most of it is fair enough, though I must say not helping me to any novel considerations.

Two paragraphs, however, contained opinions which
Mr. —— is at perfect liberty to entertain, but not, I
think, to express to me.

The one is, that I shaped what I had to say at Edin-burgh with a view of stirring up the prejudices of the Scotch Presbyterians (imagine how many Presbyterians I had in my audiences !) against Comte.

The other is the concluding paragraph, in which Mr. —— recommends me to "*read Comte*," clearly im-plying that I have criticised Comte without reading him.

You will know how far I am likely to have committed either of the immoralities thus laid to my charge.

At any rate, I do not think I care to enter into more direct relations with any one who so heedlessly and un-justifiably assumes me to be guilty of them.   Therefore I shall content myself with acknowledging the receipt of Mr. ——'s letter through you.—Ever yours faithfully,

T. H. HUXLEY.

JERMYN STREET, *March* 17, 1869.

MY DEAR DARWIN—After I had sent my letter to you the other day I thought how stupid I had been not to put in a slip of paper to say it was meant for ——'s edification.

I made sure that you would understand that I wished it to be sent on, and wrote it (standing on the points of my toes and with my tail up very stiff) with that end in view.

[Sketch of two dogs bristling up.]

I am getting so weary of people writing to propose controversy to me upon one point or another, that I begin to wish the article had never been written.   The fighting in itself is not particularly objectionable, but it's the waste of time.

I begin to understand your sufferings over the *Origin*. A good book is comparable to a piece of meat, and fools are as flies who swarm to it, each for the purpose of depositing and hatching his own particular maggot of an idea.—Ever yours,   T. H. HUXLEY.

A little later he wrote to Charles Kingsley, who had supported him in the controversy :—

JERMYN STREET, *April* 12, 1869.

MY DEAR KINGSLEY—Thanks for your hearty bottle-holding.

Congreve is no better than a donkey to take the line he does. I studied Comte, *Philosophie*, *Politique*, and all sixteen years ago, and having formed my judgment about him, put it into one of the pigeon-holes of my brain (about the H.[1] minor), and there let it rest till it was wanted.

You are perfectly right in saying that Comte knew nothing about physical science—it is one of the points I am going to put in evidence.

The law of the three states is mainly evolved from his own consciousness, and is only a bad way of expressing that tendency to personification which is inherent in man.

The Classification of Sciences is bosh—as Spencer has already shown.

Nothing short of madness, however, can have dictated Congreve's challenge of my admiration of Comte as a man at the end of his article. Did you ever read Littré's *Life of Comte?* I bought it when it came out a year or more ago, and I rose from its perusal with a feeling of sheer disgust and contempt for the man who could treat a noble-hearted woman who had saved his life and his reason, as Comte treated his wife.

As soon as I have time I will deal with Comte effectually, you may depend upon that. At the same time, I shall endeavour to be just to what there is (as I hold) really great and good in his clear conception of the necessity of reconstructing society from the bottom to the top "sans dieu ni roi," if I may interpret that somewhat

---

[1] The Hippocampus minor : compare p. 276.

tall phrase as meaning "with our conceptions of religion
and politics on a scientific basis."

Comte in his later days was an apostate from his own
creed ; his "nouveau grande Être suprème" being as
big a fetish as ever nigger first made and then worshipped.
—Ever yours faithfully,   T. H. HUXLEY.

It is interesting to note how he invariably submitted his writings to the criticism of his wife before
they were seen by any other eye. To her judgment
was due the toning down of many a passage which
erred by excess of vigour, and the clearing up of
phrases which would be obscure to the public. In
fact, if an essay met with her approval, he felt sure
it would not fail of its effect when published. Writing
to her from Norwich on August 23, 1868, he confesses
himself with reference to the lecture "On a Piece of
Chalk" :—

I met Grove, who edits *Macmillan*, at the soirée. He
pulled the proof of my lecture out of his pocket and said,
"Look here, there is one paragraph in your lecture I can
make neither top nor tail of. I can't understand what
it means." I looked to where his finger pointed, and
behold it was the paragraph you objected to when I read
you the lecture on the sea-shore ! I told him, and said
I should confess, however set up it might make you.

At the beginning of September he rejoined his
wife and family at Littlehampton, "a grand place
for children, because you go *up* rather than *down* into
the sea, and it is quite impossible for them to get
into mischief by falling," as he described it to his
friend Dr. Dohrn, who came down for ten days,

eagerly looking forward "to stimulating walks over stock and stone, to Tennyson, Herbert Spencer, and Harry's ringing laugh."

The latter half of the month he spent at or near Dublin, serving upon the Commission on Science and Art Instruction :—

To-day (he writes on September 16), we shall be occupied in inspecting the School of Science and the Glasnevin botanical and agricultural gardens, and to-morrow we begin the session work of examining all the Irishry who want jobs perpetrated. It is weary work, and the papers are already beginning to tell lies about us and attack us.

The rest of the year he remained in London, except the last four days of December, when he was lecturing at Newcastle, and stayed with Sir W. Armstrong at Jesmond.

To Professor Haeckel

Jan. 21, 1868.

Don't you think we did a right thing in awarding the Copley Medal to Baer last year? The old man was much pleased, and it was a comfort to me to think that we had not let him go to his grave without the highest honour we had to bestow.

I am over head and ears, as we say, in work, lecturing, giving addresses to the working men and (figurez vous!) to the clergy.[1]

---

[1] On December 12, 1867, there was a meeting of clergy at Sion College, upon the invitation and under the presidency of Dean Farrar and the Rev. W. Rogers of Bishopsgate, when the bearing of recent science upon orthodox dogma was discussed. First

In scientific work the main thing just now about which I am engaged is a revision of the Dinosauria, with an eye to the "Descendenz Theorie." The road from Reptiles to Birds is by way of Dinosauria to the Ratitæ. The bird "phylum" was struthious, and wings grew out of rudimentary forelimbs.

You see that among other things I have been reading Ernst Haeckel's *Morphologie*.

The next two letters reflect his views on the proper work to be undertaken by men of unusual scientific capacity—

JERMYN STREET, Jan. 15, 1868.

MY DEAR DOHRN—Though the most procrastinating correspondent in existence when a letter does not absolutely require an answer, I am tolerably well-behaved when something needs to be said or done immediately. And as that appears to me to be the case with your letter of the 13th which has this moment reached me, I lose no time in replying to it.

The Calcutta appointment has been in my hands as well as Turner's, and I have made two or three efforts, all of which unfortunately have proved unsuccessful, to find : (1) A man who will do for it and at the same time (2) for whom it will do. Now you fulfil the first condition admirably, but as to the second I have very great doubts.

In the first place, the climate of Calcutta is not particularly good for any one who has a tendency to dysentery, and I doubt very much if you would stand it for six months.

Huxley delivered an address ; some of the clergy present denounced any concessions as impossible ; others declared that they had long ago accepted the teachings of geology ; whereupon a candid friend inquired, "Then why don't you say so from your pulpits ? " (See *Coll. Ess.* iii. 119.)

Secondly, we have a proverb that it is not wise to use
razors to cut blocks.

The business of the man who is appointed to that
museum will be to get it into order. If he does his duty
he will give his time and attention to museum work pure
and simple, and I don't think that (especially in an Indian
climate) he has much energy left for anything else after
the day's work is done. Naming and arranging specimens
is a most admirable and useful employment, but when
you have done it is "cutting blocks," and you, my friend,
are a most indubitable razor, and I do not wish to have
your edge blunted in that fashion.

If it were necessary for you to win your own bread,
one's advice might be modified. Under such circumstances one must do things which are not entirely desirable. But for you who are your own master and have a
career before you, to bind yourself down to work six
hours a day at things you do not care about and which
others could do just as well, while you are neglecting the
things which you do care for, and which others could
not do so well, would, I think, be amazingly unwise.

Liberavi animam ! don't tell my Indian friends I have
dissuaded you, but on my conscience I could give no
other advice.

We have to thank you three times over. In the first
place for a portrait which has taken its place among those
of our other friends ; secondly, for the great pleasure you
gave my little daughter Jessie, by the books you so kindly
sent ; and thirdly, for Fanny Lewald's autobiography,
which arrived a few days ago.

Jessie is meditating a letter of thanks (a serious undertaking), and when it is sent the mother will have a word
to say for herself.

In the middle of October scarlet fever broke out
among my children, and they have all had it in succession,
except Jessie, who took it seven years ago. The last
convalescent is now well, but we had the disease in the

house nearly three months, and have been like lepers, cut off from all communication with our neighbours for that time.

We have had a great deal of anxiety, and my wife has been pretty nearly worn out with nursing day and night; but by great good fortune " the happy family " has escaped all permanent injury, and you might hear as much laughter in the house as at Swanage.

Will you be so kind as to thank Professor Gegenbaur for a paper on the development of the vertebral column of Lepidosteum I have just received from him. He has been writing about the process of ossification and the " deck-knochen " question, but I cannot make out exactly where. Could you let me know ?

I am anxious for the *Arthropoden Werk*, but I expect to gasp when it comes.

Turn to p. 380 of the new edition of our friend Kolliker's *Handbuch*, and you will find that though a view which I took of the " organon adamantinae " some twelve or fourteen years ago, and which Kolliker has up to this time repudiated, turns out, and is now admitted by him, to be perfectly correct, yet "that I was not acquainted with the facts that would justify the conclusion." Really, if I had time I could be angry.

Pray remember me most kindly to Haeckel, to all whose enemies I wish confusion, and believe me, ever yours faithfully,   T. H. HUXLEY.

P.S.—I have read a hundred pages or so of Fanny Lewald's 1st Bd., and am delighted with her insight into child-life.

Tyndall was resigning his lectureship at the School of Mines—

JERMYN STREET, *June* 10, 1868.

MY DEAR TYNDALL—All I can say is, I am heartily sorry.

If you feel that your lectures here interfere with your

original work, I should not be a true friend either to
science or yourself if I said a word against your leaving
us.

But for all that I am and shall remain very sorry.—
Ever yours very sincerely,   T. H. HUXLEY.

If you recommend——, of course I shall be very glad
to support him in any way I can.   But at present I am
rather disposed to d—-n any one who occupies your
place.

The following extract is from a letter to Haeckel
(November 13, 1868), with reference to the proposed
translation of his *Morphologie* by the Ray Society : —

We shall at once look out for a good translator of the
text, as the job will be a long and a tough one.   My wife
(who sends her best wishes and congratulations on your
fatherhood) will do the bits of Goethe's poetry, and I will
look after the prose citations.

Next as to the text itself.   The council were a little
alarmed at the bulk of the book, and it is of the utmost
importance that it should be condensed to the uttermost.

Furthermore, English propriety had taken fright at
rumours touching the aggressive heterodoxy of some
passages.   (We do not much mind heterodoxy here, if it
does not openly proclaim itself as such.)

And on both these points I had not only to give very
distinct assurances, such as I thought your letters had
entitled me to give ; but in a certain sense to become
myself responsible for your behaving yourself like a good
boy !

If I had not known you and understood your nature
and disposition as I fancy I do, I should not have allowed
myself to be put in this position ; but I have implicit
faith in your doing what is wise and right, and so making
it tenable.

There is not the slightest desire to make you mutilate your book or leave out anything which you conceive to be absolutely essential ; and I on my part should certainly not think of asking you to make any alteration which would not in my judgment improve the book quite irrespectively of the tastes of the British public.

[Alterations are suggested.] But I stop. By this time you will be swearing at me for attacking all your favourite bits. Let me know what you think about these matters.

I congratulate you and Madame Haeckel heartily on the birth of your boy. Children work a greater metamorphosis in men than any other condition of life. They ripen one wonderfully and make life ten times better worth having than it was.

26 ABBEY PLACE, *Nov.* 15, 1868.

MY DEAR DARWIN—You are always the bienvenu, and we shall be right glad to see you on Sunday morning.

We breakfast at 8.30, and the decks are clear before nine. I would offer you breakfast, but I know it does not suit you to come out unfed ; and besides you would abuse the opportunity to demoralise Harry.[1]—Ever yours faithfully.   T. H. HUXLEY.

An undated note to Darwin belongs to the very end of this year, or to the beginning of the next :—

The two volumes of the new book have just reached me. My best thanks for them ; and if you can only send me a little time for reading them within the next three months you will heighten the obligation twenty-fold. I wish I had either two heads or a body that needed no rest !

---

[1] This small boy of nearly four was a great favourite of Darwin's. When we children were all staying at Down about this time, Darwin himself would come in upon us at dinner, and patting him on the head, utter what has become a household word amongst us, "Make yourself at home, and take large mouthfuls."

CHAPTER XXIII

1869

IN 1869 Huxley published five paleontological papers, chiefly upon the Dinosaurs (see letter above to Haeckel, January 21, 1868). His physiological researches upon the development of parts of the skull are represented by a paper for the Zoological Society, while the *Introduction to the Classification of Animals* was a reprint this year of the substance of six lectures in the first part of the lectures on *Elementary Comparative Anatomy* (1864), which were out of print, but still in demand by students.

As President of the Ethnological Society, he delivered an inaugural address "On the Ethnology and Archæology of India," on March 9, and another "On the Ethnology and Archæology of North America," on April 13. As president of the Society, moreover, he urged upon the Government the advisability of forming a systematic series of photographs of the various races comprehended in the British Empire, and was officially called upon to offer suggestions for carrying out the project. This appears to be an amplification of Sir Joseph Fayrer's plan in 1866, with respect to all the tribes of India (see p. 394, and Appendix I.)

On April 7 he delivered his "Scientific Education : Notes of an After-Dinner Speech" before the Philomathic Society at Liverpool (*Coll. Ess.* iii. 111), one part of which deals with the attitude of the clergy towards physical science, and expresses the necessary antagonism between science and Roman Catholic doctrine, which appears more forcibly in one of his speeches at the School Board in 1871 (see ii. p. 40).

In this and other educational addresses he had suggested that one of the best ways of imparting to children a preliminary knowledge of the phenomena of nature would be a course of what the Germans call "Erdkunde," or general information about the world we live in. It should reach from our simplest everyday observations to wide generalisations of physical science; and should supply a background for the study of history. To this he gave the name "Physiography," a name which he believed to be original, until in 1877 his attention was called to the fact that a *Physiographie* had been published in Paris thirty years before.

The idea was no new one with him. Part of his preliminary lectures at the School of Mines had been devoted to something of the kind for the last dozen years; he had served on the Committee of the British Association, appointed in 1866 as the result of a paper by the present Dean Farrar, then a

Harrow master, "On the teaching of Science in the Public Schools,"[1] to report upon the whole question. Moreover, in consultation with Dr. Tyndall, he had drawn up a scheme in the winter 1868-69, for the science teaching in the International College, on the Council of which they both were.

Seven yearly grades were arranged in this scheme, proceeding from the simplest account of the phenomena of nature taught chiefly by object lessons, largely through the elements of Physics and Botany, Chemistry and Human Physiology—all illustrated with practical demonstrations—to more advanced work in these subjects, as well as in Social Science, which embraced not only the theory of commerce and government, but the Natural History of Man up to the point at which Ethnology and Archæology touch history.

It is interesting to note that the framers of this report thought it necessary to point out that one master could not teach all these subjects.

In the three later stages the boys might follow alternative lines of study according to their tastes and capacities; but of the earlier part, which was to be obligatory upon all, the report says :—These four years' study, if properly employed by the teachers, will constitute a complete preparatory scientific course.   However slight the knowledge of details conferred, a wise teacher of any of these subjects will be able to make that teaching thorough; and to

[1] See p. 401.

give the scholar a notion of the methods and of the ideas which he will meet with in his further progress in all branches of physical science.

In fact, the fundamental principle was to begin with Observational Science, facts collected; to proceed to Classificatory Science, facts arranged; and to end with Inductive Science, facts reasoned upon and laws deduced.

While he was much occupied with the theoretical and practical difficulties of such a scheme of science teaching for general use, he was asked by his friend, the Rev. W. Rogers of Bishopsgate, if he would not deliver a course of lectures on elementary science to boys of the schools in which the latter was interested.

He finally accepted in the following letter, and as the result, delivered twelve lectures week by week from April to June to a large audience at the London Institution in Finsbury Circus, lectures not easily forgotten by the children who listened to them nor by their elders :—

JERMYN STREET, *Feby.* 5, 1869.

MY DEAR ROGERS—Upon due reflection I am not indisposed to undertake the course of lessons we talked about the other day, though they will cost me a good deal of trouble in various ways, and at a time of the year when I am getting to the end of my tether and don't much like trouble.

But the scheme is too completely in harmony with what (in conjunction with Tyndall and others) I have been trying to bring about in schools in general—not to render it a great temptation to me to try to get it into practical shape.

All I have to stipulate is that we shall have a clear understanding on the part of the boys and teachers that the discourses are to [be] *Lessons* and not talkee-talkee lectures I should like it to be understood that the boys are to take notes and to be examined at the end of the course. Of course I cannot undertake to be examiner, but the schools might make some arrangement on this point.

You see my great object is to set going something which can be worked in every school in the country in a thorough and effectual way, and set an example of the manner in which I think this sort of introduction to science ought to be managed.

Unless this can be done I would rather not embark in a project which will involve much labour, worry, and interruption to my regular line of work.

I met Mr. [illegible] last night, and discussed the subject briefly with him.—Ever yours very faithfully,

T. H. HUXLEY.

I enclose a sort of rough programme of the kind of thing I mean, cut up from a project of instruction for a school about which I am now busy. The managers might like to see it. But I shall be glad to have it returned.

These lectures were repeated in November at South Kensington Museum, as the first part of a threefold course to women on the elements of physical science, and the *Times* reporter naïvely remarks that under the rather alarming name of Physiography, many of the audience were no doubt surprised to hear an exceedingly simple and lucid description of a river-basin. Want of leisure prevented him from bringing out the lectures in book form until November 1877. When it did appear, however, the book, like his other popular works, had a wide sale, and became the forerunner of an immense number of school-books on the subject.

As President of the Geological Society, he delivered an address (*Coll. Ess.* viii. 305), at the anniversary meeting, February 19, upon the "Geological Reform" demanded by the considerations advanced by the physicists, as to the age of the earth and the duration of life upon it. From the point of view of biology he was ready to accept the limits suggested, provided that the premisses of Sir William Thomson's [1] argument were shown to be perfectly reliable; but he pointed out a number of considerations which might profoundly modify the results of the isolated causes adduced; and uttered a warning against the possible degradation of "a proper reverence for mathematical certainty" into "a superstitious respect for all arguments arrived at by process of mathematics." [2]

At the close of the year, as his own period of office came to an end, it was necessary to select a new president of the Geological. He strongly urged Professor (afterwards Sir Joseph) Prestwich to stand, and when the latter consented, a few weeks, by the way, before his marriage was to take place, replied :—

JERMYN STREET, *Dec.* 16, 1869.

MY DEAR PRESTWICH—Many thanks for your letter. Your consent to become our President for the next period

---

[1] Now Lord Kelvin.   [2] See *Coll. Ess.* viii. Introd. p. 8.

will give as unfeigned satisfaction to the whole body of the Society as it does to me and your other personal friends.

I have looked upon the affair as settled since our last talk, and a very great relief it has been to my mind.

There is no doubt public-dinner speaking (and indeed all public speaking) is nervous work. I funk horribly, though I never get the least credit for it. But it is like swimming, the worst of it is in the first plunge; and after you have taken your "header" it's not so bad (just like matrimony, by the way; only don't be so mean as to go and tell a certain lady I said so, because I want to stand well in her books).

Of course you may command me in all ways in which I can possibly be of use. But as one of the chiefs of the Society, and personally and scientifically popular with the whole body, you start with an immense advantage over me, and will find no difficulties before you.

We will consider this business formally settled, and I shall speak of it officially.—Ever yours very faithfully,

T. H. HUXLEY.

I cannot place the following letter to Matthew Arnold with certainty, but it must have been written about this period.[1] Every one will sympathise with the situation :—

26 ABBEY PLACE, *July* 8.

MY DEAR ARNOLD—Look at Bishop Wilson on the sin of covetousness and then inspect your umbrella stand. You will there see a beautiful brown smooth-handled umbrella which is *not* your property.

Think of what the excellent prelate would have advised and bring it with you next time you come to the club. The porter will take care of it for me.—Ever yours faithfully,　　　　　　　　　　　　T. H. HUXLEY.

---

[1] The most probable date being 1869, for on July 1 of that year he dined with Matthew Arnold at Harrow.

The following letter shows how paleontological work was continually pouring in upon him :—

JERMYN STREET, *May* 7, 1869.

MY DEAR DARWIN—Do you recollect recommending[1] that the *Nassau*, which sailed under Capt. Mayne's command for Magellan's Straits some years ago, should explore a fossiliferous deposit at the Gallegos River ?

They visited the place the other day, as you will see by Cunningham's letter which I enclose, and got some fossils which are now in my hands.

The skull to which Cunningham refers, consists of little more than the jaws, but luckily nearly all the teeth are in place, and prove it to be an entirely new ungulate mammal with teeth in uninterrupted series like *Anoplotherium*, about as big as a small horse.

What a wonderful assemblage of beasts there seems to have been in South America !   I suspect if we could find them all they would make the classification of the Mammalia into a horrid mess.—Ever yours faithfully,

T. H. HUXLEY.

And on July 16, 1869, he writes again to Darwin :—

To tell you the truth, what with fossils, Ethnology and the great question of " Darwinismus " which is such a worry to us all, I have lost sight of the collectors and naturalists " by grace of the dredge," almost as completely as you have.

Indeed, the pressure was so great that he resolved to give up the Hunterian Lectures at the College of Surgeons, as he had already given up the Fullerian

[1] See p. 398.

Professorship at the Royal Institution.   So he writes
to Professor (afterwards Sir William) Flower:—

JERMYN STREET *June* 7, 1869.
Private, Confidential, Particular.

MY DEAR FLOWER—I have written to Quain[1] to tell
him that I do not propose to be put in nomination for
the Hunterian Chair this year.   I really cannot stand it
with the British Association hanging over my head.   So
make thy shoulders ready for the gown, and practise the
goose-step in order to march properly behind the mace,
and I will come and hear your inaugural.—Ever yours,
T. H. HUXLEY.

The meeting of the Association to which he refers
took place at Exeter, and he writes of it to Darwin
(September 28):—

As usual, your abominable heresies were the means of
getting me into all sorts of hot water at the Association.
Three parsons set upon you, and if you were the most
malicious of men you could not have wished them to
have made greater fools of themselves than they did.
They got considerably chaffed, and that was all they were
worth.[2]

And to Tyndall, whom an accident had kept in
Switzerland :—

After a sharp fight for Edinburgh, Liverpool was
adopted as the place of meeting for the Association of

---

[1] President of the Royal College of Surgeons.
[2] It is perhaps scarcely worth while exhuming these long-
forgotten arguments in their entirety ; but any one curious enough
to consult the report of the meeting preserved in the files of the
Academy, will find, among other things, an entirely novel theory
as to the relation of the Cherubim to terrestrial creation.

1870, and I am to be President; although the *Times* says
that my best friends tremble for me. (I hope you are
not among that particular lot of my best friends.)

I think we shall have a good meeting, and you know
you are pledged to give a lecture, even if you come with
your leg in a sling.

The foundation of the Metaphysical Society in 1869
was not without interest as a sign of the times. As
in the new birth of thought which put a period to
the Middle Ages, so in the Victorian Renaissance, a
vast intellectual ferment had taken immediate shape
in a fierce struggle with long-established orthodoxy
But whereas Luther displaced Erasmus, and the
earlier reformers fought out the quarrel with the
weapons of the theologian rather than those of the
Humanist, the latter-day reformation was based upon
the extension of the domain of positive science, upon
the force of historical criticism, and the sudden reorganisation of accumulated knowledge in the light
of a physical theory adequate to explain it.

These new facts and the new or re-vivified theories
based upon them, remained to be reckoned with after
the first storm of denunciation had passed by, and
the meeting at Sion College in 1867[1] showed that
some at least of the English clergy besides Colenso
and Stanley wished to understand the real meaning
of the new movement. Although the wider effect of
the scientific revival in modifying theological doctrine
was not yet fully apparent, the irreconcilables grew

[1] See p. 436.

fewer and less noisy, while the injustice of their attempts to stifle the new doctrine and to ostracise its supporters became more glaring.

Thus among the supporters of the old order of thought, there was one section more or less ready to learn of the new. Another, seeing that the doctrines of which they were firmly convinced were thrust aside by the rapid advance of the new school, thought, as men not unnaturally think in the like situation, that the latter did not duly weigh what was said on their side. Hence this section eagerly entered into the proposal to found a society which should bring together men of diverse views, and effect, as they hoped, by personal discussion of the great questions at issue, in the manner and with the machinery of the learned societies, a *rapprochement* unattainable by written debate.

The scheme was first propounded by Mr. James Knowles, then editor of the *Contemporary Review*, now of the *Nineteenth Century*, in conversation with Tennyson and Professor Pritchard (Savilian Professor of Astronomy at Oxford).

Thus the Society came to be composed of men of the most opposite ways of thinking and of very various occupations in life. The largest group was that of churchmen :—ecclesiastical dignitaries such as Thomson, the Archbishop of York, Ellicott, Bishop of Gloucester and Bristol, and Dean Alford ; staunch laymen such as Mr. Gladstone, Lord Selborne, and the Duke of Argyll ; while the liberal school was represented by Dean Stanley, F. D. Maurice, and Mark Pattison. Three distinguished converts from the English Church championed Roman Catholic doctrine—Cardinal Manning, Father Dalgairns, and W. G. Ward, while Unitarianism claimed Dr. James Martineau. At the opposite pole, in antagonism to Christian theology and theism generally, stood Professor W. K. Clifford, whose youthful brilliancy was destined to be cut short by an untimely death. Positivism was represented by Mr. Frederic Harrison; and Agnosticism by such men of science or letters as Huxley and Tyndall, Mr. John Morley, and Mr. Leslie Stephen.

Something was gained, too, by the variety of callings followed by the different members. While there were professional students of philosophy, like Prof. Henry Sidgwick or Sir Alexander Grant, the Principal of Edinburgh University, in some the technical knowledge of philosophy was overlaid by studies in history or letters; in others, by the practical experience of the law or politics; in others, again, medicine or biology supplied a powerful psychological instrument. This fact tended to keep the discussions in touch with reality on many sides.

There was Tennyson, for instance, the only poet who thoroughly understood the movement of modern science, a stately but silent member; Mr. Ruskin, J. A. Froude, Shadworth Hodgson, R. H. Hutton of the *Spectator*, James Hinton, and the well-known essayist, W. R. Greg; Sir James Fitzjames Stephen, Sir F.

Pollock, Robert Lowe (Lord Sherbrooke), Sir M. E.
Grant Duff, and Lord Arthur Russell; Sir John
Lubbock, Dr. W. B. Carpenter, Sir William Gull,
and Sir Andrew Clark.

Of contemporary thinkers of the first rank, neither
John Stuart Mill nor Mr. Herbert Spencer joined
the Society. The letter of the former declining the
invitation to join (given in the *Life of W. G. Ward,*
p. 299) is extremely characteristic. He considers the
object of the projectors very laudable, "but it is very
doubtful whether it will be realised in practice."
The undoubted advantages of oral discussion on such
questions are, he continues, best realised if undertaken in the manner of the Socratic dialogue, between
one and one; but less so in a mixed assembly. He
therefore did not think himself justified in joining
the Society at the expense of other occupations for
which his time was already engaged. And he concludes by defending himself against the charge of
not paying fair attention to the arguments of his
opponents.

It followed from the composition of the Society
that the papers read were less commonly upon
technical questions of metaphysics, such as "Matter
and Force" or "The Relation of Will to Thought,"
than upon those of more vivid moral or religious
interest, such as "What is Death?" "The Theory of
a Soul," "The Ethics of Belief," or "Is God Unknowable?" in which wide scope was given to the emotions
as well as the intellect of each disputant.

The method of the Society was for the paper to be printed and circulated among the members before the meeting, so that their main criticisms were ready in advance. The discussions took place after a dinner at which many of the members would appear; and if the more formal debates were not more effectual than predicted by J. S. Mill, the informal discussions, almost conversations, at smaller meetings, and the free course of talk at the dinner-table, did something to realise the primary objects of the Society. The personal *rapprochement* took place, but not philosophic compromise or conversion. Whether or not the tone adopted after this period by the clerical party at large was affected by the better understanding on the part of their representatives in the Metaphysical Society of the true aims of their opponents and the honest and substantial difficulties which stood in the way of reunion, it is true that the violent denunciations of the sixties decreased in number and intensity; the right to free expression of reasoned opinion on serious fact was tacitly acknowledged; and, being less attacked, Huxley himself began to be regarded in the light of a teacher rather than an iconoclast. The question began to be not whether such opinions are wicked, but whether from the point of view of scientific method they are irrefragably true.

The net philosophical result of the Society's work was to distinguish the essential and the unessential differences between the opposite parties; the latter were to a great extent cleared up; but the former remained all the more clearly defined in logical nakedness for the removal of the side issues and the personal idiosyncrasies which often obscured the main issues. Indeed, when this point was reached by both parties, when the origins and consequences of the fundamental principles on either side had been fully discussed and mutual misunderstandings removed to the utmost, so that only the fundamentals themselves remained in debate, there was nothing left to be done. The Society, in fact, as Huxley expressed it, "died of too much love."

Indeed, it is to be noticed that, despite the strong antagonism of principle and deductions from principle which existed among the members, the rule of mutual toleration was well kept. The state of feeling after ten years' open struggle seemed likely to produce active collision between representatives of the opposing schools at close quarters. "We all thought it would be a case of Kilkenny cats," said Huxley many years afterwards. "Hats and coats would be left in the hall, but there would be no owners left to put them on again." But only one flash of the sort was elicited. One of the speakers at an early meeting insisted on the necessity of avoiding anything like moral disapprobation in the debates. There was a pause; then W. G. Ward said: "While acquiescing in this condition as a general rule, I think it cannot be expected that Christian thinkers shall give no sign of the horror with which they would view the spread of such extreme opinions as those advocated by Mr. Huxley." Another pause; then Huxley, thus challenged, replied: "As Dr. Ward has spoken, I must in fairness say that it will be very difficult for me to conceal my feeling as to the intellectual degradation which would come of the general acceptance of such views as Dr. Ward holds."[1]

No amount of argument could have been more effectual in supporting the claim for mutual toleration than these two speeches, and thenceforward such forms of criticism were conspicuous by their absence. And where honesty of conviction was patent, mutual toleration was often replaced by personal esteem and regard. "Charity, brotherly love," writes Huxley, "were the chief traits of the Society. We all expended so much charity, that, had it been money, we should every one have been bankrupt."

The special part played in the Society by Huxley was to show that many of the axioms of current speculation are far from being axiomatic, and that dogmatic assertion on some of the cardinal points of metaphysic is unwarranted by the evidence of fact. To find these seeming axioms set aside as unproven, was, it appears from his *Life*, disconcerting to such members of the Society as Cardinal Manning, whose arguments depended on the unquestioned acceptance of them. It was no doubt the observation of a similar attitude of mind in Mr. Gladstone towards metaphysical problems which provoked Huxley to reply,

[1] *Life of W. G. Ward*, by Wilfred Ward, p. 309.

when asked whether Mr. Gladstone was an expert metaphysician—"An expert in metaphysics? He does not know the meaning of the word."

In addition to his share in the discussions, Huxley contributed three papers to the Society. The first, read November 17, 1869, was on "The views of Hume, Kant, and Whately on the logical basis of the doctrine of the Immortality of the Soul," showing that these thinkers agreed in holding that no such basis is given by reasoning, apart, for instance, from revelation. A summary of the argument appears in the essay on Hume (*Coll. Ess.* vi. 201, *sq.*).

On November 8, 1870, he read a paper, "Has a Frog a Soul? and if so, of what Nature is that Soul?" Experiment shows that a frog deprived of consciousness and volition by the removal of the front part of its brain will, under the action of various stimuli, perform many acts which can only be called purposive, such as moving to recover its balance when the board on which it stands is inclined, or scratching where it is made uncomfortable, or croaking when pressed in a particular spot. If its spinal cord be severed, the lower limbs, disconnected from the brain, will also perform actions of this kind. The question arises, Is the frog entirely a soulless automaton, performing all its actions directly in response to external stimuli, only more perfectly and with more delicate adjust ment when its brain remains intact, or is its soul distributed along its spinal marrow, so that it can be divided into two parts independent of one another?

The professed metaphysician might perhaps tend to regard such considerations as irrelevant; but if the starting-point of metaphysics is to be found in psychology, psychology itself depends to no small extent upon physiology. This question, however, Huxley did not pretend to solve. In the existing state of knowledge he believed it to be insoluble. But he thought it was not without its bearing upon the supposed relations of soul and body in the human subject, and should serve to give pause to current theories on the matter.

His third paper, read January 11, 1876, was on the "Evidence of the Miracle of the Resurrection," [1] in which he argued that there was no valid evidence of actual death having taken place. His rejection of the miraculous had led to an invitation from some of his opponents in the Society to write a paper on a definite miracle, and explain his reasons for not accepting it. His choice of subject was due to two reasons: firstly, it was a cardinal instance; secondly, it was a miracle not worked by Christ himself, and therefore a discussion of its genuineness could offer no suggestion of personal fraud, and hence would avoid inflicting gratuitous pain upon believers in it.

This certainty that there exist many questions at present insoluble, upon which it is intellectually, and indeed morally wrong to assert that we have real knowledge, had long been with him, but, although he had earned abundant odium by openly resisting

[1] See ii. 196.

the claims of dogmatic authority, he had not been compelled to define his philosophical position until he entered the Metaphysical Society. How he came to enrich the English language with the name "Agnostic" is explained in his article "Agnosticism" (*Coll. Ess.* v. pp. 237-239). After describing how it came about that his mind " steadily gravitated towards the conclusions of Hume and Kant," so well stated by the latter as follows :—

> The greatest and perhaps the sole use of all philosophy of pure reason is, after all, merely negative, since it serves not as an organon for the enlargement (of knowledge), but as a discipline for its delimitation : and, instead of discovering truth, has only the modest merit of preventing error :—

he proceeds—

> When I reached intellectual maturity, and began to ask myself whether I was an atheist, a theist, or a pantheist ; a materialist or an idealist ; a Christian or a freethinker ; I found that the more I learned and reflected the less ready was the answer ; until, at last, I came to the conclusion that I had neither art nor part with any of these denominations, except the last. The one thing in which most of these good people were agreed was the one thing in which I differed from them. They were quite sure they had attained a certain "gnosis' —had, more or less successfully, solved the problem of existence ; while I was quite sure I had not, and had a pretty strong conviction that the problem was insoluble. And, with Hume and Kant on my side, I could not think myself presumptuous in holding fast by that opinion. . . .
> This was my situation when I had the good fortune to find a place among the members of that remarkable

confraternity of antagonists, long since deceased, but of
green and pious memory, the Metaphysical Society.
Every variety of philosophical and theological opinion
was represented there, and expressed itself with entire
openness ; most of my colleagues were *-ists* of one sort or
another ; and, however kind and friendly they might be,
I, the man without a rag of a label to cover himself with,
could not fail to have some of the uneasy feelings which
must have beset the historical fox when, after leaving the
trap in which his tail remained, he presented himself to
his normally elongated companions.   So I took thought,
and invented what I conceived to be the appropriate title
of "agnostic."   It came into my head as suggestively anti-
thetic to the "gnostic" of Church history, who professed
to know so much about the very things of which I was
ignorant ; and I took the earliest opportunity of parading
it at our Society, to show that I, too, had a tail, like the
other foxes.   To my great satisfaction, the term took ; and
when the *Spectator* had stood godfather to it, any suspicion in the minds of respectable people that a knowledge of its parentage might have awakened was, of course,
completely lulled.

As for the dialectical powers he displayed in the
debates, it was generally acknowledged that in this,
as well as in the power of conducting a debate, he
shared the pre-eminence with W. G. Ward.   Indeed,
a proposal was made that the perpetual presidency in
alternate years should be vested in these two ; but
time and health forbade.

His part in the debates is thus described in a letter
to me from Professor Henry Sidgwick :—

DEAR MR. HUXLEY—I became a member of the Metaphysical Society, I think, at its first meeting, in 1869 ;

and, though my engagements in Cambridge did not allow
me to attend regularly, I retain a very distinct recollection
of the part taken by your father in the debates at which
we were present together.   There were several members
of the Society with whose philosophical views I had, on
the whole, more sympathy ; but there was certainly no
one to whom I found it more pleasant and more instructive to listen.   Indeed I soon came to the conclusion that
there was only one other member of our Society who
could be placed on a par with him as a debater, on the
subjects discussed at our meetings ; and that was, curiously
enough, a man of the most diametrically opposite opinion
—W. G. Ward, the well-known advocate of Ultra-
montanism.   Ward was by training, and perhaps by
nature, more of a dialectician ; but your father was unrivalled in the clearness, precision, succinctness, and point
of his statements, in his complete and ready grasp of his
own system of philosophical thought, and the quickness
and versatility with which his thought at once assumed
the right attitude of defence against any argument coming
from any quarter.   I used to think that while others of
us could perhaps find, on the spur of the moment, *an*
answer more or less effective to some unexpected attack,
your father seemed always able to find *the* answer—I
mean the answer that it was reasonable to give, consistently with his general view, and much the same
answer that he would have given if he had been allowed
the fullest time for deliberation.

The general tone of the Metaphysical Society was one
of extreme consideration for the feelings of opponents,
and your father's speaking formed no exception to the
general harmony.   At the same time I seem to remember
him as the most combative of all the speakers who took
a leading part in the debates.   His habit of never wasting words, and the edge naturally given to his remarks by
his genius for clear and effective statement, partly account
for this impression ; still I used to think that he liked

fighting, and occasionally liked to give play to his sarcastic humour—though always strictly within the limits imposed by courtesy. I remember that on one occasion when I had read to the Society an essay on the "Incoherence of Empiricism," I looked forward with some little anxiety to his criticisms; and when they came, I felt that my anxiety had not been superfluous; he "went for" the weak points of my argument in half a dozen trenchant sentences, of which I shall not forget the impression. It was hard hitting, though perfectly courteous and fair.

I wish I could remember what he said, but the memory of all the words uttered in these debates has now vanished from my mind, though I recall vividly the general impression that I have tried briefly to put down.—Believe me yours very truly,   HENRY SIDGWICK.

END OF VOL. I

Printed by R. & R. CLARK, LIMITED, *Edinburgh.*

9 781108 040457